Communication in Construction

CIS

challenges. This
ion tends to be
r short periods
ms that develop
nter-disciplinary
communication
noise/distraction
arty to another.
om interpersonal
cation between
nmunication are
tion in a more
tes the diversity
d environments.
ation problems,

Loughborough
internationally
the construction
journals. He is
lor and Francis.

d in the Scott
Ie has authored
nt organisation
ct management

the Department
of Architecture at the University of Strathclyde. He is co-editor of several books
including *Construction Industry Reports 1944–1998* (Blackwell, 2003) and the
RIBA Handbook of Construction Project Management (RIBA, 2004).

Also available from Taylor & Francis

Human Resource Management in
Construction Projects
Martin Loosemore
Andrew Dainty, Helen Lingard   Pb: 0–415–26164–3
                                                Hb: 0–415–26163–5

Spon Press

Risk Managament in Projects, 2nd Edition
Martin Loosemore, John Raftery, Charles Reilly
and David Higgon   Pb: 0–415–26056–6
                                                Hb: 0–415–26055–8

Taylor & Francis

Occupational Health and Safety in Construction
Project Management
Helen Lingard, Steve Rowlinson   Hb: 0–419–26210–5

Spon Press

Construction Safety Management Systems
Edited by Steve Rowlinson   Hb: 0–415–30063–0

Spon Press

Information and ordering details

For price availability and ordering vist our website **www.tandf.co.uk**

Alternatively our books are available from all good bookshops.

Communication in Construction

Theory and practice

Andrew Dainty, David Moore
and Michael Murray

Taylor & Francis
Taylor & Francis Group

LONDON AND NEW YORK

First published 2006
by Taylor & Francis
2 Park Square, Milton Park, Abingdon, Oxon OX14 4RN

Simultaneously published in the USA and Canada
by Taylor & Francis
270 Madison Ave, New York, NY 10016

Taylor & Francis is an imprint of the Taylor & Francis Group

Typeset in Sabon by
Newgen Imaging Systems (P) Ltd, Chennai, India
Printed and bound in Great Britain by
The Cromwell Press, Trowbridge, Wiltshire

British Library Cataloguing in Publication Data
A catalogue record for this book is available
from the British Library

Library of Congress Cataloging in Publication Data
Dainty, Andrew.
    Communication in construction : theory and practice /
Andy Dainty, David Moore, and Mike Murray.
        p. cm.
    Includes bibliographical references and index.
    1. Communication in the building trades. 2. Building – Superintendence.
3. Construction industry – Communication systems. 4. Communication
of technical information. I. Moore, David. II. Murray, Mike, 1964– III. Title.
TH215.D35 2005
690'.68–dc22   2005015291

ISBN10: 0–415–32722–9   ISBN13: 9–78–0–415–32722–0 (hbk)
ISBN10: 0–415–32723–7   ISBN13: 9–78–0–415–32723–7 (pbk)

Contents

Figures

Tables

About the authors

Andrew Dainty is a Senior Lecturer in Construction Management at Loughborough University's Department of Civil and Building Engineering. An internationally renowned researcher in the field of human resource management in the construction industry, he has published widely in both academic and industry journals. He holds a number of research grants from the EPSRC, ESRC and various government and European agencies, as well as advising a wide range of contracting and consultancy firms on human and organisational issues. He is co-author of *HRM in Construction Projects*, also published by Taylor and Francis.

David Moore is a Reader in Construction Management, based in the Scott Sutherland School of The Robert Gordon University, Aberdeen. His research activity covers areas ranging from buildability, through sustainable design and the use of solar technologies, and to perception and cognition in a construction industry context. He has been involved in award-winning research examining the behaviours of superior performing project managers, and has authored and contributed to various books covering project management organisation structures, building production management and construction project management for architects.

Michael Murray is a Lecturer in construction management within the Department of Architecture at the University of Strathclyde. He has lectured at three Scottish universities (The Robert Gordon University, Heriot Watt and Strathclyde) and has delivered research papers to academics and practitioners at symposiums and workshops in the United Kingdom and overseas. He began his career in the construction industry with an apprenticeship in the building services sector and was later to lecture in this topic at several further education colleges before becoming an academic. He is co-editor of several textbooks including *Construction Industry Reports 1944–1998* (Blackwell, 2003) and the *RIBA Handbook of Construction Project Management* (RIBA, 2004).

Preface

I believe that social study should begin with careful observation of what may be described as communication; that is, the capacity of an individual to communicate his feelings and ideas to another, the capacity of groups to communicate effectively and intimately with each other. This problem is, beyond all reasonable doubt, the outstanding defect that civilization is facing today.

(Elton Mayo, 1945)

The construction industry is wholly reliant upon effective communication between individuals, teams and organisations. However, in a project-based industry, interaction tends to be characterised by unfamiliar groups of people coming together for short periods before disbanding to work on other endeavours. This temporal dimension complicates an already problematic communication environment in which technical language, an adversarial culture and noise/distraction all combine to prevent straightforward information flow from one party to the other. Indeed, the sheer number of stakeholders involved in the processes undertaken during a construction project renders communication networks exceptionally complex and subject to change. Furthermore, with the current imperative to improve industry performance by designing and constructing more rapidly, many processes that are reliant upon effective communication occur concurrently. This increases the probability of problems occurring in the transmission and reception of vital information to the construction effort.

Texts abound on how to communicate more effectively. These proffer a variety of approaches and techniques for ensuring that information flows are well managed, that messages are conveyed properly and that the receiver interprets and acts on information in a manner that is congruent with the desired objective. So why is there a need for a book which specifically examines communication practices in the construction project environment? The answer to this question is that the complexity and dynamism of the industry's project-based structure and culture threaten to undermine the applicability of many central tenets of effective communication practice that have been applied successfully in other sectors. Indeed, many of the management practices that have evolved in response to the structural and cultural conditions typical of the sector have done little to engender an open communication environment that ensures conjoined team working, process integration and improved performance. Thus, it is incumbent upon construction managers and professionals to adopt appropriate communication strategies which accord with the particular constraints under which they work.

All too often, communication is paid scant attention in project management texts, in which it is often relegated to an 'underpinning' or 'crosscutting' issue, implicit in what managers do, but unworthy of special attention. Even in practice, it appears to be an area deemed suitable for an occasional CPD event, rather than being seen as the fundamental enabler of all processes, activities and behaviour comprising project management. In this book, the profile of communication as an enabler of effective construction project management is elevated to the top of the agenda. Rather than viewing effective communication as an important facet of project-based management, it is viewed as the essential *prerequisite* to successful project-based management; if we communicate more effectively, then other managerial processes should work more effectively as a result. Thus, by exploring the principles of effective communication and applying these concepts between people, groups, organisations and corporations, the aim is to provide a concise, but nevertheless targeted framework of theoretical and practical tools for the reader to apply as their individual situation dictates. Moreover, we hope that by grounding the art of communication within everyday project and organisational practices, this will help those working in the industry to understand more about both their own approach and that of others to the act of communication.

Rather than prescribe a set of guiding principles for managers to adhere to and thereby improve communication performance, in this book we recognise the fact that there is no single communication paradigm or panacea for a multifarious industry such as construction; managers should tailor their approaches to the situation and demands of the project in hand. Thus, we adopt a 'contingency view' in explaining the principles of effective communication. We encourage the reader to apply the principles contained within this text to identify the communication needs and difficulties within their own workplace environment. In this way they can establish more effective ways of identifying and responding to their own communication strengths and weaknesses and to the changing circumstances which characterise the dynamic construction project environment. The case studies contained within the text provide some tangible examples of how others have successfully (or unsuccessfully) developed their own communication strategies in the past, and we encourage the reader to identify their own illustrations from which appropriate methods of communication can be discerned.

Andrew Dainty
David Moore
Michael Murray
April, 2005

Acknowledgements

We would like to express our thanks to the countless people who have supplied information, discussed issues of relevance to this text and have given their valuable comments on earlier drafts of this book. We would like to express our special thanks to the individuals and companies who provided us with information for the case studies which are such an integral part of this book, particularly

Paul Cookson and Sheila Judd (Simons Group)
Alan Smith (HBG Construction Ltd)
Mike Ward (Loughborough University).

Part I

Communication concepts and contexts

This introductory part outlines the theory and practice of communication in general terms and relates this to the challenges of the contemporary construction project environment. Guidance is provided for using the book as a study aid and as a practicable handbook of communication techniques and methods.

Chapter 1

Introduction

Although managers in different industries and sectors undertake diverse tasks and activities, it has long been recognised that they spend most of their time involved in communication (Baguley, 1994: 3; Huczynski and Buchanan, 2001: 178). If viewed as a fundamentally social activity, communication activities can include engaging in conversations, listening to colleagues, networking, collecting information, directing subordinates, writing letters or transferring information through electronic devices such as telephones or computers. Hence, in many ways the communication effectivity of managers defines their performances as managers; superior performance demands superior communication. This introductory chapter defines the concept of communication and its importance in the context of the contemporary construction industry. It outlines the role and importance of the project manager in the communication process and discusses the way in which the issues central to effective communication are explored within this text. Thus, it provides a contextual backdrop to the ways in which communication will be explored in relation to the construction industry in the remaining chapters.

Defining communication

Communication can be viewed as a metaphorical 'pipeline' along which information is transferred from one person to another (Axley, 1984). It is the lifeblood of any system of human interaction as without it, no meaningful or coherent activity can take place (Thomason, 1988: 400). Nonetheless, defining 'communication' is difficult as it is such a multidimensional and nebulous concept. It can have a variety of different meanings, contexts, forms and impacts and so will mean different things to dissimilar people in different situations. This is certainly the case within the construction industry, where a plethora of different communication occurs concurrently as the following simple examples illustrate:

Example 1 Consider a personal conversation between an architect and a project manager around how a particular design detail should be constructed on site. This could comprise a focused conversation between the two as they seek to achieve a consensus on the implications of the detail for the production process, and also of any necessary changes to the detail in order that it interfaces the elements of the structure already completed. This consensus-reaching process may be facilitated by a variety of communication-enhancing tools and forms of information, such as a visual representation of the design detail. In this example, the communication process involves two specialists utilising their common understanding of industry-specific terminology and concepts through verbal and non-verbal channels. This common understanding will have taken both parties several years to achieve as they moved from novice to expert status through their experience of working in the industry. To an industry outsider (or an inexperienced insider) who does not possess such an understanding, the communication between these experts would be largely impenetrable and they would be able to extract little or no meaning from the message being communicated.

Example 2   Consider the communication of the image of the construction industry as a whole portrayed by a newspaper article about problems associated with 'cowboy builders'. Such a view may have been compounded by the typical image of the industry as invariably portrayed by the press, where it is seen as being characterised by hazardous, dirty working conditions and a 'macho' culture. In this example, it is unlikely that there will be industry-specific terminology used as the audience is the general public, rather than industry practitioners themselves. In addition, the communication is purely visual (written and perhaps graphical, but with no verbal input and so the visual aspect of body language is not part of the message). Finally, the process does not readily allow the development of a debate and/or consensus between all those 'involved' in the communication, so any errors in the message cannot be easily corrected.

Although both of the scenarios are examples of communication, each comprises radically different forms and contexts. Thus, it is important to recognise that the term 'communication' is in itself a multifarious and complex term, which can mean different things in different contexts and situations. This is a theme that will recur throughout this book as communication barriers are explored in relation to the situations within which they are rooted and techniques for improving communication are examined in the contexts for which they are appropriate.

Despite the difficulties inherent in describing what is meant by communication, it is important that a working definition of the concept is developed to underpin the analysis of communications practice contained within this book. In order to achieve this, it is appropriate to break down the term so as to define its composite dimensions. Important characteristics of the communication concept can be summarised as follows:

- Communication usually involves the transfer of *information*, a generic term that embraces meanings such as knowledge, processed data, skills and technology (Cheng *et al.*, 2001). Within construction, information is exceptionally diverse given the huge number of parties involved with construction operations.
- To communicate is to *bridge a distance* of some description, which can range from being short and simple (e.g. between two people) to long and complex (e.g. across the world) (see Skyttner, 1998). Again, in construction the disparate location of many of those involved with projects regularly necessitate communication over longer distances than in, for example, manufacturing environments.
- Successful communication (at an interpersonal level in any case) is a *social skill* involving the effective interaction between people (Hargie, 1986). Despite development in off-site production techniques, construction remains a labour-intensive industry and hence, social activity demanding communication between a wide variety of participants.
- Interpersonal communications between people usually involve *conveying facts, feelings, values and opinions* (Kakabadase *et al.*, 1988). Hence, interpersonal communications can be considered subjective and value-laden. In many respects construction is not an exact science and as such demands a degree of subjective interpretation from those participating.
- Communications do not only occur between individuals, but *can occur between groups or organisations* (Baguley, 1994). Construction is inherently a team activity involving the concurrent involvement of many specialists in order to successfully deliver project objectives.
- Communication can be seen as a *transactional process* where something is exchanged between the parties involved (see Eisenberg and Goodall, 1993: 22). Construction can be seen as a series of transactions between the parties involved. Facilitating these transactions has been widely recognised as a key issue for the industry to address if it is to improve its performance in the future.

These wide-ranging perspectives on communication all suggest that communication is essentially about the transfer of information between people. Thus, the point of communication in most cases is that one person (or team or organisation etc.) wishes another to receive information from another. Within an organisational context this could be to convey an instruction to influence the actions/behaviour of others, or may involve an exchange of or request for information. To some extent, this interaction will be determined by the rules and norms of social behaviour, as it is people who translate the meanings and utilise the information (Gayeski, 1993).

This also suggests that communication has to be a two-way process, as unless the transmitter of the information receives feedback that the message has been received, then they will be unsure as to whether communication has actually taken place or if it has taken place successfully. Put simply therefore, communication involves the giving out of messages from one person and the receiving (and successful understanding) of messages by another in response (Torrington and Hall, 1998: 112). The ways in which these messages can be conveyed are multifarious and may include speech, body language, writing, graphical or electronic media or any combination of these forms. As such, communication can be viewed as a professional practice where appropriate rules and tools can be applied in order enhance the utility of the information communicated, as much as it can a social process of interaction between people.

The importance of effective communication

The importance of effective communication to individuals, teams and organisations cannot be overstated. Virtually every text on how to manage people will contain important principles of how to communicate effectively with the workforce. At an individual and team level, people find it difficult to function in the industry if they do not develop a mutually agreed communication *modus operandi* to underpin their work activities. Similarly, the management of organisational processes also demands that robust and effective communication channels are developed which enable their various components to be conjoined appropriately. The importance of communication to organisations is succinctly summarised by Armstrong (2001: 807):

- *Achieving coordinated results* – organisations function by means of the collective actions of people, but independent actions lead to outcomes incongruent with organisational objectives. Coordinated outcomes therefore demand effective communications.
- *Managing change* – most organisations are subject to continuous change. This, in turn, affects their employees. Acceptance of and willingness to embrace change is likely only if the reasons for this change are well communicated.
- *Motivating employees* – the degree to which an individual is motivated to work effectively for their organisation is dependent upon the responsibility they have and the scope for achievement afforded by their role. Feelings in this regard will depend upon the quality of communications from senior managers within their organisation.
- *Understanding the needs of the workforce* – for organisations to be able to respond effectively to the needs of their employees, it is vital that they develop an efficient channel of communication. This two-way

channel must allow for feedback from the workforce on organisational policy in a way that encourages an open and honest dialogue between employees at all levels and the top-level managers of the organisation.

Given the benefits of effective communication outlined above, the corollary of poor communications for an organisation is that employees will misread management decisions or react to them in a way that was not intended. Similarly, managers will misunderstand the needs of employees and will therefore suffer from lower performance and a higher turnover of staff. Communication has become even more important as the business world has begun to shift towards what is now described as a 'knowledge-based economy'. A knowledge revolution has underpinned the shift towards a predominance of service sector organisations (and a gradual erosion of the manufacturing base). Arguably, most large construction firms have now become service sector firms, outsourcing the majority of their productive capability and effectively acting as managers of the process. Professional and managerial employees dominate their payroll and it is these 'knowledge workers' whose intellectual capital becomes the substance which underpins organisational growth and development. The challenge for such organisations is how to engender knowledge sharing and nurture 'communities of practice' for improved performance. Communities of practice are where groups of people who share a concern for the same issues, or set of problems, come together and interact on an ongoing basis (see Wenger *et al.*, 2002: 4). They can be considered the building blocks of effective problem-solving within contemporary organisations.

Another requirement for effective communication in construction stems from the industry's propensity to undergo change and transition. Coping with change is more problematic in traditional industries like construction, which have shown a reluctance to embrace new ways of working, but is arguably more important, considering the disparate pools of knowledge that must be combined within construction projects. In the past, a 'silo' like mentality has prevailed which has been shown to stymie knowledge sharing within the industry (Dainty *et al.*, 2004). However, effective communication has the power to break down such barriers by bringing people together, thereby propagating improved collaboration and integrated working within the sector. Thus, effective communication can be seen as the cornerstone of future industry improvement.

Why study the human aspects of communication in construction?

As will be explored in more depth in Chapter 2, the construction project environment presents a particularly problematic arena within which to apply communication practices proven to be effective in other sectors.

Everyone involved in construction plays a part in a complex communication network. Seeing the project environment as an interconnected network of actors is appropriate because every such venture, no matter how small or well defined, can be successfully completed without interactions and transactions between people and organisations. This is best appreciated by way of a simple example. Consider the construction of a simple, single floor domestic extension to a traditional brick-built dwelling. The communication network for even a relatively straightforward project such as this will involve interaction, between the house owner (the client), the designer (the architect), the contractor (and its workforce), the materials suppliers and the planning and building control officers as a minimum. Now consider the construction of a new terminal building and runway at a major international airport. This type of project might require years of planning and investigation before the onsite activities can even be mobilised. It will probably involve many hundreds of organisations and thousands of individuals from around the world working together in an integrated and collaborative effort. The information flows involved with such an endeavour are potentially enormous, and yet must be carefully managed if the desired outcomes are to be achieved. However, in both examples, the success or otherwise of the project will depend upon the effectiveness of those involved to convey their needs and perspectives to others. It is the multifarious nature of human interaction that renders the understanding of communication and effective communication practice so problematic.

Construction is not peculiar in its reliance upon effective communication. Indeed, without the ability to communicate, it is possible that any contemporary organisation would cease to exist as we understand them (Katz and Kahn, 1978). Construction does, however, present a particularly complex (and, for that matter, interesting) environment within which to explore communications phenomena. Because it is project-based, its groups and networks are temporary in nature and relationships and interactions continually change to reflect the dynamic nature of the workplace. The projects themselves can be summarised in terms of their uniqueness, complexity and discontinuity. Every project will have unique characteristics and will involve a number of different actors, all of whom will have a specific and finite involvement with the endeavour. Overcoming the complex and temporal constraints that projects place on their participants is fundamental to their successful development (Goczol and Scoubeau, 2003). Hence, whilst construction is an established sector within which processes and protocols have been refined to facilitate communication within it, an element of uncertainty always exists that has the potential to undermine the communication channels necessary for project success.

Despite the massive investment in information and communications technologies (ICT) in recent years, it is impossible to divorce interpersonal and inter-group communication from the construction process. It is in these interactions that the success or failure of any project is rooted, and not the speed of an Internet connection or the compatibility of two Computer Aided Design (CAD) systems, important as these technical issues are to underpinning the process. Thus, as was alluded to above, it is essential that construction is viewed as a social activity within which communication plays a vital role.

Given the centrality of effective human communication to the success of the industry, it is surprising that so many of the recent calls for the industry to improve its performance concentrate on process and product improvements at the expense of the need to improve the complex inter-organisational and interpersonal relationships that define the industry's culture (see Emmitt and Gorse, 2003: 2). Indeed, people working in construction cannot be relied upon to act and interact in an identical manner because they will be coming from a variety of different perspectives and backgrounds and thus, will have differing needs from their interactions with others. As was discussed earlier, it is precisely because of people's idiosyncrasies and the diversity of the industry that construction presents such a fascinating environment within which to explore communication practices, as this book will reveal later.

Why study organisational aspects of communication in construction?

Communication is often treated as a background or underpinning variable in determining organisational performance, conceptually related to the structure, culture, leadership and rewards of an organisation (Church, 1996). Such a definition belies the central importance of communication as an enabler of other organisational activity. Indeed, given its influence over the efficacy of the construction industry's processes and practices, the lack of attention to the organisational aspects of communication within many construction management texts is somewhat surprising. Although information technology solutions can help to relay information rapidly and effectively, understanding of the social, structural and cultural constraints of the organisation on the communication process is arguably more important. Thus, as well as exploring the processes of human interaction and its impact on effective communication, a concurrent emphasis of this text is to explore how human interaction takes place within the broader contextual framework of the organisation. By exploring human communication processes within both the temporary (project) and permanent (firm) organisational contexts of the industry, this allows the interplay of these mutually influential factors to be understood and a more holistic understanding of how communication can be managed more effectively in the future.

Communication within the 'knowledge economy'

Knowledge is the vital resource which lies at the heart of both organisational and project success (Nonaka and Takeuchi, 1995; Egbu, 1999). As was alluded to earlier in this chapter, within the post capitalist society, knowledge cannot be considered just another resource alongside the traditional factors of production (land, labour and capital), but is in fact the *only* meaningful resource (Drucker, 1993). Unsurprisingly, there has been a great deal of attention on developing ways to manage knowledge more effectively in recent years, primarily through the development of new ICTs. However, this emphasis has largely been at the expense of efforts to explore the human dimensions of effective knowledge management (Scarbrough, 1999; Swan *et al.*, 2000). Arguably, the overemphasis on technological solutions for managing knowledge within large organisations has contributed to the relatively high failure rate of knowledge management (KM) initiatives within many industries and organisations (Ambrosio, 2000; Carter and Scarbrough, 2001; McDermott and O'Dell, 2001). Thus, there is a practical need to integrate KM programmes with human resource management (HRM) policy, to ensure its effective contribution to the performance of the modern business (Blackler, 1995; Swan *et al.*, 2000; Carter and Scarbrough, 2001).

Explicit knowledge represents only the metaphorical 'tip of the iceberg' of the entire body of knowledge. As Polanyi (1966: 4) states 'we can know more than we can tell', which implicitly suggests that there are difficulties inherent in the communication of tacit knowledge, or even that there are aspects we as individuals cannot convey to others (Kane, 2003). Tacit knowledge is therefore not easily visible and expressible, but is highly personal and therefore difficult to communicate or share with others (Smith, 2001). This acknowledgement reflects the belief of many theorists and practitioners, that knowledge must be harnessed and communicated more effectively for organisations to develop and for performance to be improved. The importance of knowledge capture and transfer will be returned to throughout this book in relation to the changing role of construction organisations within the knowledge-based economy. Notably, many firms are transforming into service-oriented firms, which employ few people outside of those involved in their core service provision. This has created new challenges for such firms who have to procure additional services and productive capacity from external suppliers, which arguably opens up new communication interfaces which have to be managed.

The principles of effective communication

The problematic context of communicating in construction raises questions as to how the industry can go about overcoming the structural and cultural conditions and constraints which define its operation, in order that it can develop an infrastructure that facilitates more effective communication in the future. Moreover, it suggests that the industry needs to find ways of effecting change within the sector in such a way as to overcome existing cultural constraints on the sector's development. In a theoretical sense, applying methods of effective communication should be fairly straight-forward, but how theoretical perspectives actually translate in practice will depend upon their interpretation by the people that work in the sector. Arguably, those with experience of working in construction have developed skills to cope with such a challenging communication environment which enable them to overcome the inherent difficulties of short-term interaction. However, given that construction is not homogenous and involves people from a wide variety of craft, managerial and professional backgrounds, there is no guarantee that the use of espoused 'good practices' will result in successful outcomes. Indeed, the extent to which existing knowledge has revealed whether there are generally applicable performance-enhancing principles remains questionable (Marchington and Grugulis, 2000).

In his review of the conceptual foundations of organisation communica-tion theory, Deetz (2001) refers to what he terms the 'discourse of normative studies'. This perspective sees organisations as '...naturally existing objects open to description, prediction and control'. This work has implicitly viewed the world as ordered and well-integrated and has as such, ignored the impact of organisational goals or individual member positions. A more critical view of organisational life would see such a perspective as crude in that it fails to reflect the socio-cultural development of the firm and the relations which flow from it. Thus, the perspective in this book is not to prescribe a set of generic or normative 'tools for improved communication', but to raise aware-ness of the issues that must be understood if those working in construction are to communicate more effectively. It is left to the reader to decide upon the most appropriate ways to communicate based on their individual circum-stances and the nature of the power relations which surround them. For example, an appropriate approach to communication for a project manager will be radically different from a managing director because of their different situational circumstances and the nature of their role within the firm.

The belief that communication panaceas do not really exist may appear incongruent with a concurrent overarching aim of this text alluded to earlier, which is to suggest *practicable* methods for improving communications within construction projects and organisations. However, there remain fundamental guiding principles underlying improved performance that need to be applied in a way that accords with the particular project or situation at hand. In other words, the techniques used should be contingent upon the particular circumstances to which they are to be applied (see Loosemore *et al.*, 2003: 27). Taking an 'open systems' view of organisations and recog-nising the impact that environmental variables have on the decision of how to communicate most effectively is an underlying philosophy that will be adhered to throughout the book. Indeed, it is a philosophy that distinguishes this book from others that take a more prescriptive view of what defines appropriate or effective communication and/or management practice. Even the examples and suggestions that populate this text have been selected to reflect the changing nature of organisations and processes, which in turn characterise the construction industry of the twenty-first century.

The role and importance of effective communication to the project manager

Although the theories, principles and practices discussed within this book will be of interest to virtually anyone interested in the construction industry, the orientation of this book is primarily towards project managers with responsibility for overseeing the construction process. These are the key individuals who manage the production function and who effectively form the vital communication interface between the project and the wider organisation for which they work. As will be discussed in more depth later in this book, the term 'project' suggests that the endeavour in question has a uniqueness, novelty and transience not found in static production environments. This, in turn, also conveys an implied risk associated with the activity (Turner, 1998: 4). The role of the project management function is to manage the systems that relate to these features; namely the scope of work, the project organisation, the quality, the cost and the duration of the project. Communication is an essential ingredient of all of these managerial requirements.

The need for the project manager to be an effective communicator is self-evident when considering the number of occasions that communication has been cited as a primary cause of project failure. Even a cursory examination of both industry and academic literature will reveal the significance of communication in shaping unfavourable outcomes within projects of all sizes. For example, over four decades ago Higgin and Jessop (1965) wrote a seminal report which identified communication problems between project stakeholders at both business and project levels and highlighted the severe implications for project performance. In the early 1980's Guevara and Boyer (1981) further revealed the problems of information overload, gate keeping and distortion experienced in construction companies based in the United States. More recently, Boudjabeur and Skitmore (1996) found evidence of untimely, inaccurate and insufficient information in projects undertaken in the United Kingdom.

Project managers as communication facilitators

As will be explored in greater depth in Chapter 3, at the core of the communication process are two key processes: encoding and decoding (Clevenger and Matthews, 1971). Encoding refers to the process of sending of messages and encompasses constructing stimuli that represent meaning (see Norton, 1978). A signal, or message, is produced that may (or may not) be emitted by the sender (Roberts *et al.*, 1987). It encompasses the activities within a person that are involved in transforming inner thoughts, ideas, feelings and information into messages. The transfer of information can include speech, nonverbal signs and writing. Decoding, on the other hand, is the process of actively listening to messages (Norton, 1978). This involves turning sensations into meaning or patterned codes (Roberts *et al.*, 1987). Decoding involves the transformation of sensory input into significant interpretation(s). Decoding activities of communication include listening, reading and interpreting of nonverbal signs. When applied to the context of construction management, a project manager's competency in encoding and decoding arguably plays a crucial role in achieving project outcomes. This is because they are positioned between the production focused staff involved in the project and the design and client team overseeing the project's development. Given that so much project information flows through them in both directions, if they are effective in communicating their thoughts through the encoding and decoding process, they are more likely to achieve their desired outcomes and hence, be successful in their role.

Clearly, project management within construction is a very demanding role and one which requires a multitude of different skills, competencies and abilities (see Dixon, 2000). Defining competent communication behaviour has been an active area of research and application in the field of communication for over 30 years with researchers broadly defining communication competence from a variety of perspectives (Henderson, 2004). Notably, project managers must communicate both up and down the supply chain, transcending team and organisational boundaries and overcoming physical, contractual, cultural and psychological barriers along the way. An effective project manager must be able to communicate at all levels; with directors, peers, functional managers and suppliers (Turner, 1998: 437). They must also be able to communicate across a wide range of specialist disciplines at each level (such vertical and horizontal communication is sometimes light-heartedly referred to as 'helicopter' communication). In addition, they must be able to achieve these communication goals under the extreme pressures that project-based environments create, where everything is bounded by rigid time and resource constraints.

The importance of communication to project managers' leadership abilities

A vital ingredient of project management competency is leadership, of which communication is a vital component. An effective leader is able to communicate a vision which gives meaning to the work of others (Handy, 1993: 117). Thus, using communication to generate buy-in to their managerial approach towards the achievement of desired project outcomes represents the major challenge for the project manager. They must also be aware of the implications of their actions for the broader corporate communication issues which will be conveyed beyond the project environment. Recognising that no team, project or organisation exists in a vacuum is an essential precursor to establishing an effective and appropriate communication style.

The recognition of communication within project management bodies of knowledge

As has been emphasised above, a key characteristic of the construction industry is its project-based structure. In fact, construction is one of the oldest and most established project-based sectors. Newer project industries and professions include information technology, electronics manufacture, management consulting and some aspects of aerospace and marine engineering. The common facet of all of these sectors is that teams of people with complementary skills and knowledge are assembled for finite periods of time, during which they have to combine their efforts in a way that supports the objectives of the endeavour. Accordingly, the various project management bodies have established standards against which individuals can assess their own knowledge and performance, thereby setting benchmarks for performance within the profession. It is important within a book such as this that the relevant aspects of such standards are alluded to in order that the reader understands how adopting the approaches described within the text can help to shape their competence in relation to what is required of the discipline as a whole.

Existing project management competence and competency standards are contained within the bodies of knowledge developed by the various associations and institutes that oversee the discipline. Within the United Kingdom, the best known of these is probably the Association of Project Managers Body of Knowledge (APMBok) (see APM, 2000). The APM standards make specific mention of communication under their 'People' related competency area, highlighting its fundamental importance to making project management work (APM, 2000: 50). They emphasise the importance of different media: oral, body language, written (textural, numerical, graphic), paper, electronic etc., but also stress that it is the *manner* in which communication is delivered that is perhaps more important than the medium by which it is conveyed. This infers that communicating effectively forms a core facet of the project manager's role and one which should be developed through training and relevant activities. It also refers to the crucial interrelated function of information management for underpinning the communication process. Reference will be made to this important set of competence standards and the ways in which the techniques and approaches discussed can support the development of these competencies for those seeking personal development within a project management role.

The aims, focus and structure of this book

As is discussed above, the focus of this book is on the human aspects of communication in construction projects and organisations, primarily from a project management perspective. This is not to exclude the importance of informatics and technological support for communication (which is itself addressed within Chapter 8), but the effectiveness of technology is ultimately dependent upon the ways in which information is encoded, transmitted, decoded and interpreted by the people involved. Managed effectively, this process should yield the outcomes desired of those initiating the communication, but poorly managed communication can undermine the processes at stake and ultimately the performance of the team, project or organisation involved.

The book is oriented towards students of construction related subjects such as construction/project management, engineering, quantity surveying, architecture etc. *and* construction practitioners wishing to improve their knowledge of communication processes and practices within the sector. In this respect, this book aims to bridge the gap between the theory of communication and how it is practiced within the industry. Its focus is on the softer skills that underpin effective project management supported by theoretical insights taken from both the construction management and wider business fields. By combining established communication theories from the psychology, social psychology and management fields and relating them to construction events, it aims to provide practicable steps that those working in the industry can take to ensure an open and free-flowing communication system.

The structure of the book explores the topics discussed above within nine concise chapters divided into three broad parts as follows.

Part I: communication concepts and contexts

This introductory part outlines the theory and practice of communication in general terms and relates this to the challenges of the construction project environment.

- Chapter 1: *Introduction* – this chapter has introduced the topic of communication in construction. It has explained the crucial role of effective communication for the future development of the sector and its managers, and has outlined the principles upon which the text is founded.
- Chapter 2: *The challenges of communicating in the construction project environment* – this chapter explains the intricacies and peculiarities of the construction project environment and the impacts and constraints that this places on communicating effectively. It outlines the crucial role of the project manager in managing the communications issues that arise from the dynamic and complex nature of the sector.
- Chapter 3: *Theoretical perspectives on construction communication* – this chapter charts the evolution of communication in both academic discourse and practical management application. It introduces the

different types of communication and media that will be explored elsewhere within this book.

Part II: from individuals to corporations: communication types and techniques

This part explores the practice of communication in the industry, identifying how the constraints identified in Chapter 2 can be overcome through the effective use of communication principles, tools and techniques.

- Chapter 4: *Interpersonal communication* – this chapter examines one-to-one communication between people in the industry. It explores how personal interaction between people can be better managed in order to improve the transfer and understanding of information.
- Chapter 5: *Group and team communication* – this chapter explores group behaviour in the context of construction. It examines the uniquely complex team-based environment of the construction industry and suggests ways in which intra-group barriers to effective communication can be addressed.
- Chapter 6: *Organisational communication* – this chapter examines the issues and problems inherent in communicating within an organisational setting. Communicating between teams and projects is a problematic but necessary aspect of organisational life within the industry, particularly if knowledge is to be taken hold of, harnessed and built upon to improve the performance of construction firms. Moreover, communicating messages and conveying a desired vision and set of values can be problematic, particularly in distributed project-based environments.
- Chapter 7: *Corporate communication* – this chapter takes a broader view of communication in considering how an organisation or team can portray a positive corporate image to the outside world. It explores the ways in which the firm relates to its clients, supply chain, competitors and its own staff in a way that aligns with its strategic objectives.

Part III: future directions for construction communication

This section explores the future of communication in the construction industry. It examines the role and importance of information and communications technologies as enablers of communication both now and in the future. There are also suggestions as to how the industry might begin to address its previous failings by drawing upon the principles expounded within the book.

- Chapter 8: *Information and communications technology* – this chapter examines the ways in which information technology is revolutionising the way in which we communicate within the industry. It explores the barriers to the greater usage of ICT as an enabler of change, how the benefits of leading edge technologies can be realised and how this might change the way in which people interact and interrelate in the industry of the future.

* Chapter 9: *Conclusions and future directions in construction communication* this chapter concludes the book by examining what the industry needs to do to address its complex and dynamic project-based structure. It identifies the core communication challenges for the industry and discusses the crucial role that the project manager has to play in improving communication in the future. Finally, it suggests a set of research themes and topics, the exploration of which should inform the development of a more open and effective communications culture in the future.

The book has been written and structured such that a reader can refer to chapters with relevance to their own communication context. Where relevant, points of reference from elsewhere within the book have been highlighted in order to enable the reader to gain some underpinning material to support their understanding of the issues presented. The case studies provided at the end of Chapters 2, 4, 5, 6, 7 and 8 attempt to put the issues and theories discussed into a practical context in order that their relevance to the way in which communication can be improved within the industry can be better appreciated. The case studies may be of additional interest in relation to many of the general themes and topic explored within the text as well as in relation to the specific chapter context to which they relate.

Summary

This chapter has defined communication in the context of construction and, moreover, has discussed its central importance to the effectiveness of construction projects and organisations. It has suggested that there are no panaceas for overcoming the inherent difficulties of such a complex and fragmented industry, but that managers must tailor specific communication approaches in accordance with the unique contexts of the projects that they manage and the organisations for which they work. Thus, this text rejects the normative view of communication that it merely comprises a set of tools and techniques that can be generically applied to any circumstance. Rather, it sees communication as a contextually bound phenomenon and one which demands bespoke approaches to the situations that present themselves.

The aims of this book are to explore the issues determining the effectiveness of communication within the industry, and to identify the guiding principles and practices that could enable managers to devise strategies for overcoming these constraints. The underlying premise of this text is that by applying theory to facilitate the understanding of communication within the industry and combining this with a sound understanding of the practical realities of communicating within the sector, impediments to effective communication can be ameliorated. The focus of this book is on the human aspects of communication rather than technological facilitation of the communication process. This is because exclusively focusing upon the technological aspects of communication effectively ignores the way in which individuals and groups encode, send, receive and interpret information, and crucially, the role of the project manager in ensuring an effective communication environment. By raising the profile of human communication as a crucial enabler of industry improvement, this book aims to ensure that the issue takes it rightful place as the primary management consideration within the sector in the future.

Critical discussion questions

1 Discuss the role that effective communication has to play in improving project performance and meeting client needs in the contemporary construction industry.
2 Consider an unsuccessful project with which you are familiar. Evaluate the extent to which poor communication practices played a part in the failure of the project team to deliver on its objectives.

Chapter 2

The challenges of communicating in the construction project environment

Along with shipbuilding and aerospace, the construction industry is one of the oldest and most established project-based sectors (Keegan and Turner, 2003). Indeed, although the last 100 years has seen the industry evolve from a predominantly craft-based industry to a multifaceted service, production and manufacturing sector, its basic unit of operation has remained largely unchanged; the project. While construction has adopted 'new' and innovative initiatives (such as the increasing utilisation of plant and off-site 'volumetric' construction) to varying degrees, its evolution is ultimately determined by the constraints of site-based production. Because the industry has largely continued to construct in situ, the project-based approach has remained one of bringing together a diverse collection of craft, professional and managerial staff to disparate locations. These people will work collectively for short periods of time, before they disband to work on other ventures (see Bryman *et al.*, 1987).

Within the temporary organisation of the project, participants have a range of objectives, not all of which will be complementary in nature. Competing needs and objectives naturally lead to feelings of discord and tension, which in turn raise the possibility of conflict within the construction project team. Each of these states may, in turn, stymie communication and delimit the achievement of project objectives. Thus, the maintenance of effective communications is an effective way of ensuring that project teams are working together effectively and ultimately, to ensuring the successful delivery of projects.

Before techniques for effective communication are explored, it is important to understand the difficult context in which managers operate. Within project-based industries such as construction, barriers to effective communication are complex and multifarious because of the number of actors which govern the success of construction practices. Accordingly, this chapter explores how the nature of the construction industry and its project-based structure constrains effective communication within it. It outlines how the nature of the industry's production system leads to short-term interactions between the parties involved in the delivery of projects, and how these can, in turn, undermine the development of effective communication structures. It also explores how the physical, contractual, cultural and temporal nature of construction projects act as 'noise' that interferes with the efficacy of communication processes within the industry. This chapter therefore acts as a contextual backdrop to the remainder of the book in which methods for enabling more effective communications and overcoming these barriers and constraints are explored.

The nature of project-based working

For many years, poor communication practices have been recognised as a serious delimiting factor within the construction industry. Within the United Kingdom, a succession of government-commissioned reports has berated the industry for its apparent inability to communicate effectively, both internally and externally (Emmerson, 1962; Banwell, 1964; Latham, 1994; Egan, 1998; Strategic Forum, 2002). Over the past two decades, an abundance of management texts have expounded more effective communications mechanisms as a route to the idealised concept of the 'post-modern organisation'. However, there is a paucity of evidence as to the take-up of such approaches within the construction industry. Rather, construction appears to have ignored many of the espoused communication panaceas for the 'knowledge-based' economy. Indeed, despite the prevalence of reports and rhetorical commentary surrounding the need for improving communication performance, there has actually been little (if any) noticeable enhancement in communication performance in recent years (Emmitt and Gorse, 2003: 16). This raises questions as to why the modern construction industry appears unable to communicate effectively and moreover, what barriers to communication must be overcome by the techniques advocated within this book.

The characteristics of projects

As was discussed earlier, first and foremost, the construction industry is a *project-based* industry. A project is a temporary endeavour undertaken to create a unique product or service (Project Management Institute, 2000). This infers that every project will provide a differing set of challenges and contexts which must be carefully managed in order to achieve a successful outcome. Turner (1998: 3) embodies this when he defines a project as:

> an endeavour in which human, financial and material resources are organized in a novel way to undertake a unique scope of work, of given specification, within constraints of cost and time, so as to achieve beneficial change defined by quantitative and qualitative objectives.

Deconstructing Turner's definition reveals several aspects of projects which present particular difficulties from both a communication and project management perspective. These include:

- *The organisation of human, financial and material resources* – the project manager must reconcile a range of disparate and finite resources in order to achieve the desired objectives. Whereas some of these resources are inanimate objects, people are very unpredictable and as such, require different skills and competencies if they are to be managed effectively.
- *Novel forms of organisation* – every project will demand a different approach to its coordination and management, which in turn will demand bespoke communication approaches tailored to the internal and external stakeholder needs of the situation at hand. Large and complex projects involving multiple stakeholders will demand a more sophisticated structure than a simple endeavour involving a few people.
- *A unique scope of work and specification* – given the unique nature of construction projects (even standard 'off the shelf' buildings will face a unique combination of site constraints, stakeholder attitudes and opinions and performance criteria), this demands that project managers generate a bespoke communication strategy for every project.
- *Constraints of cost and time* – it is an inevitable reality that all project-based working is constrained by time and cost limitations. This requires that the project manager develops a communication strategy which conveys the required information rapidly and precisely in order to ensure that timely and cost appropriate actions are undertaken throughout the project.
- *Quantitative and qualitative objectives* – whereas quantitative objectives are measurable and therefore easy to determine performance against, softer performance criteria such as quality and relationship-based measures are more subjective and hence, open to interpretation by project stakeholders. Nevertheless, it remains incumbent upon the project manager to meet stakeholder needs and to communicate this effectively to them.

Equally important, from a communications perspective at least, is that projects are characterised by transience; patterns of short-term involvement which inevitably delimit opportunities for teams to establish firm and permanent communication channels which accord with the needs of all of the stakeholders involved. The impermanent nature of the project team demands that participants put in place written and unwritten protocols to ensure that information flows are managed effectively until such a point that a bespoke communication *modus operandi* can be established. Such systems, which are often embodied within the clauses of construction contracts, can constrain the natural development of relationships. If they were allowed to evolve naturally they could enhance the performance of projects as they are progressively elaborated (or allowed to proceed steadily from the outset in order for objectives to be met – see Project Management Institute, 2000). As such, there is little time to establish and embed communications channels and processes. They must be rapidly established to suit both the needs and complexity of the project *and* the individual needs of the project participants.

Considering that the objectives of the project will be variously defined according to the specific perspective of the individual stakeholder, it stands to reason that communication needs will also differ depending upon who is involved in the process. Moreover, there is evidence to show that external stakeholders to the core project team find it difficult to penetrate project boundaries. Loosemore (2000) showed that on some projects, the pressures, cohesion, loyalties, focus and momentum that can develop become so intense that the construction project team effectively seals itself off from the outside world. This isolationism can be extremely damaging to the ability to set up the flexible communication channels and processes necessary for coping with the change that transcends project-based working.

The characteristics of construction projects, people and processes

Construction project activity is extremely diverse, ranging from simple housing developments to highly complex infrastructure projects. However, all types of construction project, regardless of their size and complexity, have some common characteristics, which warrant elaboration considering their impact on the communication process. Some of these features render the industry more complex and difficult to manage than even other project-based sectors. These features are summarised by Loosemore *et al.* (2003) as:

- *Their unique, one-off nature* – unlike other project-based sectors, where prototypes can be tested before production gets underway, construction projects tend to be one-off, unique undertakings that are designed and constructed to meet a particular client's product and service needs. Even where a standard design is being used, individual sites will present their own individual challenges with regards to ground conditions, logistical constraints and the prevailing weather, not to mention the fact that different parties and their behaviours will provide a completely different set of relationships and interrelationships for every venture undertaken. The inter-dependence of project participants in this regard is particularly significant. This can lead to significant risks and stresses for people working on a project, which arise from learning-curve problems associated with new work activities and ever changing workplace relationships.

- *Their tendency to be awarded at short notice* – many construction projects are awarded following a period of competitive tendering, where possibilities for thorough planning are often limited. Even where non-competitive methods are used, there is usually pressure to deliver projects as rapidly as possible, often requiring concurrent design and construction activities which place project participants under considerable strain. Having been awarded a contract, a design consultancy or contractor has to mobilise a project team comprising an appropriate blend of skills and abilities to meet the project demands quickly. As such, construction organisations need to respond to sudden changes in workload as there can be no guarantee of how much work will be undertaken at any particular time (Hillebrandt and Canon, 1990).

- *The labour intensiveness of construction activity* – the construction industry remains one of the most people-reliant industrial sectors; staffing costs represent the majority of costs on most projects. The industry employs an extremely diverse range of people from a wide range of occupational cultures and backgrounds, including people in unskilled, craft, managerial, professional and administrative positions, and these diverse groups of employees operate as an itinerant labour force, working in teams to complete short-term project objectives in a variety of workplace settings. Hence, the industry is made up of many disparate organisations which come together in pursuit of shared project objectives, but also individual organisational goals. These are not necessarily compatible or mutually supportive and they might not align with peoples' personal objectives, which can lead to competing demands on those working within project-based environments. These features make construction one of the most challenging environments in which to manage people effectively, and thereby ensure that they contribute to organisational success.

- *Jargon, semantics and the potential for misunderstandings* – the industry's fragmented structure and culture and its technical nature have led to the creation of both formal and informal languages developing around its processes and people. For example, the vocabulary and glossaries used by project management practitioners can lead to communication problems (Delisle and Oslon, 2004). A lack of standardisation in size, quality and commonality of meaning in glossaries that are targeted at project management practitioners mean that individuals define similar processes differently. The use of technical language and jargon (written and spoken) are common within many occupations. In construction this may be evident in drawings, specifications, method statements and other project documentation. Verbal communication may also be ineffective and time wasting when noise occurs in a transfer due to the use of overly complex language. The use of particular terms may be intended to create confusion or embarrass the receiver or can be used to communicate an occupational identity. Construction lawyers arguably do this by the copious use of Latin, thereby securing a sense of mystique. Craft operatives may use slang terms for tools and components as a passage through which apprentices are socialised into their occupation.

- *The reliance on a mobile workforce* – construction projects are, for the most part, constructed in situ. Even with the increased use of off-site fabrication and the wider use of pre-fabricated components, the final product is normally assembled and completed in the required site location. This necessitates the employment of an itinerant workforce, which can move from one project location to the next. This transience of location and the temporal nature of project teams poses many problems for workers such as longer working days, more expense in travelling to work and managing work/life balance issues, since their families may not be as mobile. Transience also arises *within* projects, since the composition of teams normally changes during different project stages. The complexity brought about by the dynamic and transient nature of projects is compounded by the fact that construction involves people from many organisations, backgrounds and locations (see later).

- *An ingrained male-dominated culture* – construction is one of the most male dominated industries in virtually every developed society. Men dominate both craft trades and professional and managerial positions within the sector. This reliance on male employment leads to many challenges, such as skills shortages caused by recruiting from only a portion of the population, difficulties in the management of equal opportunities and workforce diversity and considerable challenges in terms of creating an accommodating atmosphere in which individuals' diverse skills and competencies are fully utilised (Dainty *et al.*, 2000a,b). It also leads to a particular type of person being attracted to work within the sector. The homogeneity that this promotes has arguably contributed to the development of an intransigent workforce unwilling to embrace change. It may also have determined the ways in which communication takes place.

- *An increasingly diverse labour market* – notwithstanding the homogeneity which is a characteristic of the construction workforce, there is an increasing tendency and indeed, skills-shortage induced necessity, for the industry to begin to draw upon more diverse sources of labour. Within the United Kingdom, this is manifesting itself in an influx of foreign labour, particularly in the South East of the country. The recent expansion of the European Union, with accession being granted to several new states have enabled more skilled workers to enter the United Kingdom to offset immediate skills needs. These workers are almost certainly to bring with them different languages and cultural norms which will further complicate the construction communication challenge. The project manager may find himself/herself having to bridge a cultural language divide which is particularly significant in terms of ensuring a safe and healthy work environment for all of those engaged in on-site activities.

These unique challenges posed by construction activities require companies to balance project requirements with their organisational priorities and individual employee expectations. As will be explored throughout this book, it is the industry's inability to manage the communication process surrounding this challenge which has caused many of the enduring problems that plague the industry today. Poor communications practices and failing to overcome the types of barriers discussed above can have dire consequences for construction projects, particularly in large and highly complex schemes. One obvious implication is the detrimental financial consequences which stem from an inability of participants to communicate effectively. However, in a physical and inherently dangerous industry like construction, of greater consequence is the impact that poor communications practices can have on health and safety. A good example of this is the 1981 Hyatt Regency Hotel disaster in Kansas City, where 114 people were killed and around 200 others injured when two crowded walkways collapsed onto a crowded dance floor. Wearne (1999) provides a detailed account of the cause of this disaster which was traced to the design and fabrication of the 36-ton walkways that were suspended above the dance floor. The design involved suspending the walkways (one beneath the other) using continuous hanger rods attached to the roof of an atrium. However, due to envisaged buildability problems, design drawings of the structural engineers were changed to incorporate a new two-rod system. This led to a discrepancy between the structural drawings and the shop drawings and, despite much iteration between the fabricator and designer, the structural engineering firm signed off the structurally inadequate two-rod design, with a disastrous box-beam connection. Wearne (ibid.) notes that the Hyatt Regency case was a troubled project that had experienced serious failures whilst in construction. Chaotic lines of communication and confused chains of responsibility between designers, architects and contractors existed which were manifested in a woefully inadequate inspection system. This type of man-made disaster is partly attributable to the multiple numbers of interactions between project stakeholders who rely on each other to fully complete design work. A more detailed example of the implications of the industry's poor communication practices is provided in the case study at the end of this chapter.

The communication imperative for construction organisations

Communication may be a taken-for-granted component of organisational life, but it is by no means straightforward in complex organisations such as large construction companies. The Building Industry Communications Research Project (BICRP, 1966) was instigated by the National Joint Consultative Council of Architects, Quantity Surveyors and Builders (NJCC) in 1966 (see Boyd and Wild, 2003). It created a new set of theories about complex, interdependent and uncertain situations which have informed our understanding of communication within the sector. These acknowledge the informal nature of communications and reveal the variety and confusion inherent in the industry's activities. In particular, a complex interdependence was found between the unstable roles and relationships that exist within the industry as a result of its project-based structure. This instability is at the heart of the difficulties that construction organisations still face today.

The dynamic patterns of involvement that characterise operations within the sector effectively present a changing set of relationships for all of those involved with projects. It is these different social constructions that make effective communication problematic within complex organisations (Thompson et al., 1990). Some organisations in more static industries attempt to minimise the problems this creates by appropriate recruitment, selection, training and socialisation of their staff (Smallman and Weir, 1999), or perhaps by distorting and/or re-shaping information in order that it meets organisational needs (Bella, 1987). However, this cannot mitigate the difficulties of so many independent social actors interacting across many internal and external interfaces, as is the case in construction. Managing this process represents the central communication challenge for construction project organisations and those who manage them.

Without effective communication, it would be impossible for any construction company to compete in such a challenging operating environment. In addition, many construction firms have realised that effective communication provides an important route to competitive advantage. By communicating effectively they should ensure that their own employees, and the other firms that they interface with, contribute to the achievement of organisational objectives. Effective intra-organisational communication is problematic as there will be many structural and cultural barriers to managing and transferring information across project and professional boundaries. Communicating across the entire supply chain represents an even more significant challenge given the fragmentation and sheer number of organisations involved in the project delivery process. Nonetheless, effective communication is essential for the successful delivery of performance goals (productivity, profitability and repeat working opportunities) and even more significantly, a safe and healthy workplace environment as was discussed earlier.

There are a wide variety of different reasons as to why some forms of communication fail. These include a lack of understanding (no proper means available to convey information so that it is easily understood); information overload (simultaneous and continuous submission of information that goes beyond a person's capacity to take it in and understand); and poor information quality (procedures or instructions from one person to another are of such poor quality that they are inappropriate to the situation when followed) (Williams, 1988; cf. Wantanakorn *et al.*, 1999). According to Baguley (1994: 13) the types of factors causing communication difficulties can be further refined thus:

- *A lack of clear objectives* – without a clear intention, this leads to uncertainty of the message, and to confusion between the transmitter and receiver.
- *Faulty transmission* – usually occurs because the message is sent via an inappropriate medium or channel. It can also occur when a receiver is expected to absorb too much information or when they lack an insight into the circumstances around the transmission.
- *Perception and attitude problems* – are related to misunderstood messages where transmitter and receiver attribute different meanings so that a shared understanding is not possible.
- *Environmental problems* – from distractions and noise, a lack of appropriate communications media and physical distance.
- *Chinese Whispers* – the phenomenon of a message being gradually distorted as it passes along the message chain. The longer the chain, the more distorted the message would become.

It can be appreciated that all of these barriers to communication are likely to be more prevalent in construction companies than in many other types of organisation. For example, project objectives will differ from one stakeholder to another and so information can take on a variety of different meanings. Transmission problems are commonplace in construction, such as where incompatible information technologies can lead to a failure of one party to understand the needs of another. Perception and attitudinal barriers exist whereby participants from different professional occupational back-grounds interpret situations in different ways, such as where an architect and a quantity surveyor attempt to reconcile the need for architectural functionality with the need to reduce cost. Environmental problems are particularly acute within the industry. Often, project stakeholders are based at a considerable distance from the project location and yet decisions involving many organisations must be made rapidly throughout a project's lifetime. Finally, the phenomenon of 'Chinese Whispers' is perhaps more acute in construction than in more integrated sectors, as the message chain is longer and the boundaries (or organisational interfaces) through which information must pass are more numerous. Consider the relaying of a design 'message' along a chain. This commences with the client, passing through the client representative, the design team, the contractor, sub-contractors and individual operatives carrying out the operations on site. There is the potential for distortion at each stage along this perilous journey as will be explored later in this book.

The preceding discussion has emphasised the potential barriers that can impact on communication within the industry. However, often these factors combine to inhibit the effective transfer of information within and between organisations. Thus, understanding the communication challenge within the industry demands that they are all understood in terms of the ways in which they impede either the internal or external communications of the organisation as the following sections explore.

Barriers to effective intra-organisational communication in construction

Barriers to effective communication within construction begin within the firm. With the possible exception of small and highly specialised firms (or micro-businesses), construction companies will contain a variety of different types of professional, managerial, skilled and unskilled craft and administrative employees. All of these individuals will need to communicate across their divisional, departmental, professional and hierarchical interfaces in order to successfully fulfil their specific function. Project teams usually comprise of a mix of people drawn from different functional divisions and departments who each lend particular expertise to the production effort (Langford et al., 1995). This is a classic trait of 'matrix' organisation structures within which employees report to (temporary) project and (permanent) functional line managers. The mutual reliance of the team member's on each other's skills characterises much of the industry's operation, and yet this presents a major communication challenge for construction organisations.

An example of how these functional interfaces can detrimentally affect communication within organisations is provided by research carried out Moore and Dainty (1999, 2000, 2001) who explored the nature of communication between the members of 'integrated' design and build project teams. They found that the espoused advantages of the design and build procurement route (such as integrated working, seamless communication, single points of contact etc.) could be undermined by issues arising from the rigid professional cultures of individual participants within project workgroups. These were found to inhibit the integration of the design and construction processes. They found that cultural non-interoperability stymied the abilities of the group to manage change, innovate and improve the performance of both the design and construction processes. Moreover, project responsibilities, which tend to be delineated along professional lines, provided design and construction solutions that failed to fulfil the potential of this procurement route.

Table 2.1 Differences between the jargon used by architects and mechanical and electrical engineers

Term	To an architect	To an engineer
Air-conditioning	Any cooling system – probably comfort cooling	One particular system that is cooled, heated, humidity-controlled and ducted. Not comfort cooling
Contractor	Builder	Plumber or electrician
Duct	Anything needed for hidden services	A galvanised steel air-distribution system to HVAC publication DW142
Low-energy structure	Wood-framed, lightweight structure with turfed roof	Concrete structure with exterior insulation and heat exchanger
Natural ventilation	Windows	The passive passage of air-through grilles, chimneys, stacks and exposed mass
Pipe size	The actual size of the pipe with everything else included, such as insulation	The mean diameter of the pipe itself – excluding insulation thickness

Source: Dadji (1988).

Overcoming these professional boundaries is especially significant when it is considered just how many professional roles exist within the sector. Dadji (1988) reports such phenomenon in relation to the relationship between architects and mechanical and electrical (M&E) engineers. Many architects often see M&E engineers as 'techies' who use different jargon to architects. Examples are provided in Table 2.1, which shows how misunderstandings can arise based on the different parlance used by these professionals. This shows how two construction industry professionals could interpret different meaning into the same technical phrase.

Gorse and Emmitt (2003) explored interpersonal communication during construction progress meetings. They found that management and design team interaction is subject to task-based interaction norms, but that these could be affected by outbursts of emotional interaction, which greatly influenced group behaviour. This emphasises the individualised nature of communication dynamics within the sector, and the propensity of people within the industry to act in ways which do not necessarily support interaction requirements. This is exacerbated by the large number of unexpected events that occur within construction projects. Loosemore (1996, 2000) carried out extensive research into the patterns of behaviour that occurred during crises within the industry. His research revealed the complex patterns of communication that arise in response to unforeseen events occurring during construction projects, many of which are unnecessary or counterproductive in terms of resolving the problem at hand. The findings of this research add weight to the assertion that communication within the construction firm is exceptionally complex and problematic. Ensuring that employees comply with the values of the organisation and work towards the fulfilment of their goals demands approaches to communication which can break down the interfaces and overcome behavioural norms associated with construction work.

Barriers to effective inter-organisational communication in construction

Construction projects involve multiple organisations working in an inter-disciplinary environment and the task of aligning a common objective within a temporary team is fraught with difficulty. Considering that any construction project endeavour essentially involves a change and transition from an initial concept, through to a design and specification and ultimately into a completed structure, effective communication can be seen to lie at the heart of the industry's operations. However, given the problems of communicating within firms, moving information across organisational boundaries can be seen to be exceptionally problematic. The complex nature of the industry has contributed to its extreme fragmentation relative to other production sectors. Major construction projects now involve a complex array of stakeholders, each of which has a temporary involvement within its lifecycle. Every individual effectively communicates from his/her own frame of reference, albeit moderated to some extent by an equally complex legal framework of rights and responsibilities. Considering that much of the specification, design, management and productive activity take place concurrently during projects with a resultant emphasis on rapid communication between the parties involved, this renders effective communication exceptionally problematic.

The industry has a legacy of mistrust and allocating blame when things go wrong. Informally, project stakeholders are likely to express different opinions about an event that has occurred during a project or indeed the overall success of a completed project. These 'competing narratives' have been found to be more prone to exist in projects that involve uncertainty; integration and urgency (Turner and Muller, 2003). Such conditions are common in most construction projects and may act as the catalyst for the different beliefs expressed by project participants. Boddy and Paton (2004) have examined this issue and argue that these differences originate in the subjective interpretations of a project itself and from the cultural, structural, political and career interests of its participants. Poorly managed projects that do not resolve important competing narratives will generate apathy, distrust and scepticism within project teams. Thus, a critical project management task is to understand the source of competing narratives and to manage them for the good of the project. This issue appears to have received little attention from construction researchers, although there have been some research projects concerning the culture of the sector and its impact on project success (e.g. Rooke *et al.*, 2003).

Bodensteiner (1970) revealed that if organisational problems occur or high levels of uncertainty exist, then people tend to rely on informal channels for information. Bodensteiner's research findings are pertinent in that periods of uncertainty within construction projects are very common, typically a result of unplanned incidents creating a 'project crisis'. Resolving such a crisis will normally require rapid decisions to be made by one or more parties in the project team. Unwillingness to make decisions quickly may result in uncertainty, which can, in turn, result in error or inappropriate decision-making. In a study of human errors in building projects, conducted in Sweden by Josephson and Larsson (2001), it was found that many design errors concerned the coordination of information of drawings from different groups of designers. As Turner and Pidgeon (1997) note, where a number of agencies are brought together to complete a task, there is more likelihood of communication failures occurring than when a task is completed within a single agency. This is exacerbated by the fact that the individual cultures of organisations involved in a task will each have their own distinctive subculture and version of rationality as was discussed earlier. This rationality may give rise to erroneous assumptions about the portion of the problem that is being handled by other units.

Considering that the construction supply chain is long and fragmented, there are more opportunities for problems between transmitters and receivers of information. Similarly, the more organisations that are involved, the greater the opportunity for inappropriate media, misaligned objectives and conflict and confusion that appears to comprise an everyday aspect of working within the industry. Indeed, it has been suggested that those choosing a construction career may do so because they thrive in adversarial and pressurised workplace environments (Loosemore *et al.*, 2000). This may be symptomatic, however, of the tendency for organisations to attempt to offload risk and responsibilities to other members of the supply chain, rather than work collaboratively to resolve the issues arising. Either way, this has profound effects on the communications climate within the sector.

Structural constraints on effective construction communication within construction

The structural attributes of the construction industry have also been found to inhibit communication. Of particular relevance here is the dominant sequential view of construction activities which tends to prevail. This can prevent open and effective communication between parties within the construction supply chain. A good example of this is provided by O'Cathain and Gallacher (1999) who investigated the effects of communication failures on the process of design, procurement and installation of construction façades. There were widespread communication and quality problems at all stages of the façade procurement process. They identified problems with the transmission, receipt, comprehension and application of information. The research identified several cases where architects had not supplied drawings on time or where they made continuous design changes after manufacturers, fabricators or suppliers had placed orders for glass and materials and had scheduled production runs. Rectifying this situation demanded that architects rid themselves of the old-fashioned sequential view of procurement activity and managed such issues concurrently. Another example is provided by Love and Li (2000) who conducted an in-depth investigation of two construction projects to explore the causes and cost of rework. They similarly concluded that rework was attributable to the sequential nature of the supply chain, which resulted in poor communication and decision-making processes. Thus, the implications of inflexible structures and procedures can be seen to have a detrimental impact on the performance of projects.

The role that organisational structure has in facilitating or impeding communication is another interesting discussion point. Organisational structure can be formally laid down in some respects, and left to evolve informally in others. The formal organisational chart can be considered a concrete manifestation of a set of structural variables including the differentiation of tasks. However, Dingle (1997) also acknowledges that *informal* communication is important for the running of projects, as long as it is controlled. This reinforces the earlier work of Hopper (1990) who suggested that construction project managers should allow teams to move between formal and informal structure in order to achieve their mandate. Figure 2.1 shows likely informal communication paths layered over a formal hierarchy. It shows how the

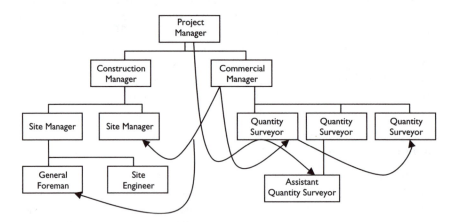

Figure 2.1 Formal and informal communication.

Source: Adapted from Dingle (1997).

organisation chart is a poor representation of interactions between people, and of who may have the power in an organisation. A single piece of communication can move through parts of the structure whilst circumventing others, thereby undermining the formal communication protocol. Thus, formal organisation charts can be either outdated, inaccurate, or both (Hage *et al.*, 1971). Organisational charts can be viewed as an 'optimistic' set of expectations about relationships and communication between employees (Dalton, 1959). This is particularly true within dynamic construction organisations, as Maurer (1992) states:

> Because organization designers are neither clairvoyant nor omniscient, they cannot construct in advance an organization structure that can anticipate all future contingencies, especially for an organization coping with an unstable and unpredictable external environment.

As discussed earlier, research undertaken by the Tavistock Institute (BICRP, 1966) was one of the first studies to document the use of informal systems within the construction process. It specifically suggested that informal procedures seem to produce more realistic phasing of decisions and more realistic flexibility in the face of what is described as the inevitable uncertainties in the construction process. The report was critical of persistent unreal assumptions attached to formal project control mechanisms, suggesting that as a result there has been an 'inappropriate application of techniques of scientific management'. It is suggested that informal systems enable projects to be completed without major delay.

The degree to which formal communication channels are defined within construction projects is commonly thought of as being dictated by a combination of procurement route and its associated contractual form. Formal project communication patterns are imposed on the project team and are therefore considered to be pre-designed, rather than evolving. However, project participants are likely to bring their own pre-conceived ideas about what they wish to communicate. These ideas will be shaped according to their previous interactions with the other team members on previous projects and how amicable those relations were. Current interest in project partnering and its inherent philosophy of cooperation would tend to suggest that the divergence between contractually prescribed and actual communication structures will continue to broaden in the future.

Challenges in communicating design information

A particularly challenging aspect of construction work is how to convey design information. It has been estimated that a delay in the supply of adequate information during the construction phase of a project may contribute 21–30 per cent to the total delay within projects (see Ganah *et al.*, 2000).

Thus, the threat of liquidated damages coerces the contractor to push ahead with production, often with insufficient production information. This problem has its roots in the separation of design and production activities and is compounded by the proliferation of specialist design input. Even where design and build (D&B) procurement is employed, the timely coordination of structural, M&E services with architectural drawings is often inadequate. The adoption of manufacturing techniques such as concurrent engineering as a mechanism to facilitate seamless design and construction (see Anumba and Evbuomwan, 1997; Kamara *et al.*, 2000) has arguably had little impact apart from for those clients with control over their projects. This emphasises the role that poor communication has had on the communication of design information. Research undertaken by Fox *et al.* (2001) has also shown that the adoption of a Design for Manufacture (DFM) philosophy within the construction industry has also been thwarted by the communication problems that prevent effective design collaboration. Thus, the majority of projects continue to suffer problems in communicating design information at all stages within the project. Indeed, the frequent use of fax machines to transmit urgent production information between design office and site testify to the chaos associated with construction industry project supply chains. Unfortunately, the risks associated with this are normally passed down the supply chain to the plethora of specialist contractors who are left to cope by making pragmatic decisions on site.

One way in which the industry has attempted to overcome design communication problems is through the increasing use of ICT within project environments (e.g. project intranets, e-mail, web cams, 3-D visualisation models, nD modelling), which have the potential to facilitate non-verbal communication. However, it is acknowledged that despite improving the speed of communication, such technology has had limited influence on either the efficiency of the process or the quality of information exchange (Alshawi and Underwood, 1999). Ganah *et al.* (2000) have shown that the uptake of computer visualisation within the construction industry is low. The most common methods used to communicate between design teams and site teams continue to be traditional methods such as 2-D drawings, face-to-face meetings, written statements, telephone and fax. This may be because of the slow take up of new technologies within the industry and the continuing problems of compatibility between different systems.

Clients who commission projects on a regular basis and who have greater direct control over their projects clearly have the potential to exert more influence on the efficiency of communication within their projects. Property developer Stanhope has, for example, piloted a new computer simulation model called Design and Logistics Integration (DALI) that allows a building's construction process to be modelled in three dimensions. The model has apparently improved communication between the designer–contractor–specialist–contractor interface by assisting in the visualisation of programming, buildability and logistics information (*Building*, 2001a). Thus, client leadership in defining communications protocols may provide an incentive for more construction consultants, contractors and specialist firms to engage with new technologies in order to ensure that they communicate more effectively in the future.

New communication challenges for the contemporary construction industry

Although the ability of the various stakeholders involved in construction activity to communicate effectively is to some extent constrained by the nature of the industry and its fragmentation, its importance is such that in recent years it has been identified as a primary cause of the industry's poor performance. Since 1994, pressure has been growing for the industry to improve its working practices in order that it can deliver better value for money for its clients, improve its own profitability and improve the way in which it treats people working within the sector (Latham, 1994; Egan, 1998; Strategic Forum, 2002). This pressure can be attributed to a range of factors, which together can be seen to present a new set of communication challenges for an industry already struggling to cope with its ingrained culture, fragmented delivery structure and apparent reluctance to embrace new ICT solutions. Relevant factors are discussed briefly below.

The performance improvement imperative

In recent years, much has been made of the industry's need to change and respond to client demands for improvements in the delivery of the sector's projects. There has been a steady increase in the quality of service and product expected by clients procuring construction work. Inevitably, improving performance demands more effective ways of communicating between parties in order that client needs are correctly interpreted and their expectations managed. The customer-focused performance rhetoric that has surrounded this debate has seemingly fuelled a revolution towards the adoption of new processes designed to radically improve the operational practices of the industry. These, in turn, demand new levels of communication between the people involved which have arguably not been taken account of in their implementation (see Green and May, 2003). Thus, although improved communication is a prerequisite of a step change in performance improvement, it could be described as a missing ingredient in industry performance improvement recipes!

The drive for process integration

Defects and rework in construction can be attributed to design errors, materials failures, and/or workmanship problems. However, Moore and

Dainty's (2001) work (see earlier) shows that communication problems often lie at the heart of all such problems. In addition, the continued use of an 'over-the-wall' separation to design and construction activities provides an ideal breeding ground for confusion and hence, communication difficulties. Indeed, Harrower (2003), Group President of the Special Engineering Contractors (SEC), argued that his member contractors and key manufacturers continue to be denied the opportunity to join design teams and therefore cannot contribute knowledge about value engineering, sustainability and whole life costs. He suggested that communication between designers on public sector contracts is inhibited by hierarchical contract structures involving sequential appointments. Harrower claims that this leads to consultants handing down designs to contractors/subcontractors who are required to redesign them, resulting in delays and disruption. An indication of this failure to engage in 'concurrent engineering' philosophy is noted in Cox and Hamilton's (1991) *Architects Handbook of Practice Management*. This offers prescriptive guidance to architects that appears to discourage closer collaboration, stating that 'consultants may be tempted to discuss technical matters direct with sub-contractors, but it should be made clear that the only valid instructions are those formally by the architect to the main contractor'. Arguably, this 'rule book' continues to be rigorously observed by too many design consultants today, resulting in a legacy of disputes that continues to thwart improved interaction in many projects. It belies the informal communication structures which, if allowed to grow and develop, can ensure the efficient working temporary organisations such as projects.

In order to overcome the problems that the separation of design and construction activities presents, recent reports have emphasised the importance of integrating processes and importantly, the various members of the supply chains involved. Perhaps the most influential of these, the Egan report (1998), defined an integrated project process in terms of several important characteristics:

- They utilise the full construction team, bringing the skills of all the participants to bear on delivering value to the client.
- They are explicit and transparent, and are therefore easily understood by both project participants and their clients.
- They lead to increased efficiency of project delivery due to elimination of the constraints imposed through the (traditionally separated) processes of planning, designing and constructing construction products.
- They promote concurrent working with regard to the input of designers, constructors and key suppliers.
- They allow teams of designers, constructors and suppliers to work together over a number of projects, thereby continuously developing and innovating the product and the supply chain.

This definition places emphasis upon the supply chain, and Egan recognises that partnering (as one method of organising relationships) does not present itself as being an easy option for either constructors or suppliers. A particular problem is that it requires recognition by those involved of the need to 'unlearn' well-established, traditional relationships such as exist in distributed teams. Indeed, the UK construction industry has a long history of formalised relationships between clients and constructors, and for a large part of that history the relationships have been formed in the context of a fragmented and boundary-ridden industry (Moore, 1996). Relationships between members of the industry have been constrained by the boundaries between professional and vocational institutes. Teams may well be distributed and still contain boundaries that represent professional/vocational limitations.

The need to develop 'high performance' teams

Construction, being an inherently team-based activity, requires that people work together, synergistically combining their skills and expertise for successful project outcomes. A team is formed for the purpose of bringing together complementary skills to achieve an outcome that could not be achieved as efficiently or as effectively by a group of individuals (Ingram *et al.*, 1997; Bragg, 1999; Nesan and Holt, 1999; Rosenthal, 2001). When the team works together effectively, this can lead to group accountability and mutual responsibility for achieving results. This forms the basis of a 'high performance' team environment. Effective communication is another key enabler of high-performance team working, in that high performance teams thrive in open communication climates, where ideas and information are freely exchanged in a collaborative workplace environment (Huczynski and Buchanan, 2001: 875). Thus, effective communication and the achievement of a high performance team-based environment can be seen to go hand-in-hand.

The changing nature of procurement practices

In recent years, there have been fundamental changes to the way in which projects are procured and moreover, the ways in which participants interact within them. This is in part a reflection of the recognition of the value of effective human and organisational communication (see Cheng *et al.*, 2001). By encouraging people to work in partnership and by developing long-term relationships through strategic alliances this arguably breaks down many of the communication barriers alluded to earlier. Nevertheless, the communication demands that new procurement practices have placed upon the sector have arguably been ignored as, although the same parties are involved, the constitution of their relationships changes fundamentally under each system. For example, the elevated position of the contractor under management forms of procurement reconstitutes the communicative relationship between themselves and the client. This results in contractors' staff requiring a completely different approach to communicating with parties throughout the project organisation structure than would be the case under more traditional procurement routes. Thus, the need to understand and manage the interpersonal and inter-organisational communication patterns that arise from these new relationships is essential to their future development.

Workforce behaviours

Organisational communication is arguably a key factor in embedding new ways of working necessary for change within construction organisations. Communication is not merely a mechanism to convey or transmit information, but is a tool by which workforce attitudes and behaviours can be challenged, manipulated and changed (see Townly, 1994). In recent years, some construction companies have begun to recognise the power of 'softer' behavioural competencies in defining the success of an organisation (Moore et al., 2002). This acknowledgement stems from the realisation that it is the behavioural input to a project's development that determines its success. Such behaviours are manifested and conveyed to others in the ways in which project participants communicate. Thus, training, developing and supporting people in improving their communication skills is central to the improved performance of the sector in the future. How such skills should be developed remains a moot point however, as will be explored later in this book.

Workforce diversification

Multiculturalism is an increasingly prominent feature of the modern construction industry in many countries. Indeed, even in the homogenous UK construction industry, pressure is growing for the sector to begin to attract more women, ethnic minorities and other groups which have been historically underrepresented within the sector (Dainty et al., 2002). This has been shown to present a serious challenge in ensuring communication between the various cultural groups (Loosemore and Lee, 2002). People from disparate cultural and social backgrounds may interpret different meanings into the communications that they have. Ensuring that this does not inhibit collaborative working and performance is bound to present a major challenge to construction companies in the future. There are also additional implications for the health, safety and welfare of workers as was alluded to earlier in relation to workers entering the labour market without English language skills (also see later in relation to globalisation of the construction industry).

Globalisation

A popular theme in contemporary organisational communication concerns the need to adapt to competitive and dynamic markets (Huczynski and Buchanan, 2001: 203). The onset of global construction markets has recently begun to be felt within the sector at large. Fast growing economies such as the Chinese construction industry offer massive market potential for European companies willing to invest in attempting to exploit them. However, taking advantage of new markets demands either the wholesale recruitment of staff from the indigenous market being targeted and/or the relocation of expatriate staff from the home country. Internationalisation has increased the intercultural communication problem for many organisations operating within many sectors (see Deresky, 1994). Indeed, experiences in construction to date have shown that many managers of construction companies who work overseas are ill-prepared for their role and experience significant cultural and communications problems with their subordinates and local management counterparts (Loosemore and Al Muslmani, 1999). Thus, providing the language skills and cultural knowledge for employees needing to communicate in unfamiliar environments, and ensuring robust communication paths for managing projects in different countries represent fundamental challenges for construction firms seeking to expand into international markets.

New communications technologies

As will be explored in depth in Chapter 8, ICT has the potential to redefine the ways in which the industry operates. However, in other respects it can create as many communication problems as it solves. For example, whilst on one hand, modern technologies such as mobile telephones have revolutionised communication within construction, they have also changed the way in which people interact, reducing the face-to-face contact that can overcome many of the relational problems known to beset the industry. Furthermore, they load additional pressure on to those working within the sector who are expected to be available to deal with problems as they arise. The onset of electronic commerce (E-commerce) and web-based technology is also potentially significant for the future development of the sector. The ways in which it might speed up procurement, design, and production processes in the future however, depends upon how effective people are at utilising it and ensuring that electronic channels of communication are compatible and aligned throughout the supply chain. A potential implication of an increase in the availability of information is the danger of information overload whereby the user cannot determine the information that is important to support decision-making. Thus, the adoption of information and communication technologies must be treated with discretion and caution if their use is not to undermine the tenets of effective human communication and interaction.

Communication challenges for the construction project manager

A recurring theme in this book is that the construction industry relies on the abilities and skills of line managers to a greater extent than do most other sectors. Pinto and Pinto (1991) showed that managers' efforts to clarify and establish shared agreement for deliverables had a positive influence on team member satisfaction. Other research has demonstrated the important role of managers' communication in ensuring job satisfaction (Henderson, 1987; Pettit *et al.*, 1997) and positive relations between co-workers and managers (Klauss and Bass, 1982). This presents a significant leadership and communication challenge for project managers who must ensure that their teams are motivated and working towards a successful outcome for the endeavour with which they are engaged. However, construction companies also require their managers of projects to make decisions that align with the overall strategic philosophy of the organisation. Moreover, they are expected to communicate the organisational vision in such a way that the remainder of the team collectively direct their efforts around them. So far, this chapter has explored communication from a primarily organisational perspective, but as was discussed in the introductory chapter, this book is aimed at those with responsibility for the management of the production function within the industry. Thus, it is appropriate that attention is now turned to those with responsibility for managing the production function.

Within construction, the project manager is one of the key individuals in the successful delivery of a project. In effect, the project manager's role is to try and balance their decisions in such a way as to reconcile the needs of all parties involved. They must bring together a group of disparate individuals, who may have never worked together before, to work collaboratively on a task that none of them may have relevant prior experience of. There is a surprising lack of research exploring the interaction of the construction manager with other key construction professionals (Emmitt and Gorse, 2003: 17). However, insights can be gleaned from studies into project management abilities in other sectors. These show that in order to be effective in bringing together the project team, a project manager must develop a common identity amongst team members; a set of values and norms which will help them to achieve their objective (Turner, 1998: 425). Such a demanding role requires dynamic leadership qualities, defined by Turner (1998: 435) as:

- *Problem-solving ability and results orientation* – effective managers are usually of above average intelligence, are able to analyse a situation and to solve complex problems and are driven to achieving the desired goals.
- *Energy and initiative* – effective project managers must be able to work under pressurised circumstances whilst at the same time taking the initiative in resolving difficulties and avoiding problems.

- *Self-assurance* – effective project managers must take resolute actions and be confident in their own abilities. This is not to say that they should act in a brash manner, but be self-assured in order to instil confidence in their own abilities amongst their team.
- *Perspective* – the effective project managers must be able to see beyond the team and to how they influence the organisation as a whole. They must be able to overview the project-level activities in a way that allows them to see discontinuities in the management systems being applied if problems are to be successfully resolved.
- *Negotiating ability* – effective project managers must overcome their lack of line authority over their functional managers through negotiation and influence. This extends to their external liaison with the project sponsor in terms of managing their expectations of the delivery of the project in accordance with their expectations.
- *Communication* – effective project managers must be able to communicate at all levels, from the project sponsors down to the junior members of their team. Interaction with different project stakeholders and team members requires different communication abilities.

Communication ability is defined as a discreet trait within Turner's analysis of the effective project manager. However, it could also be argued that it forms a cross cutting ability necessary for the achievement of all of the other factors as all of these aspects must be communicated to both team members and external stakeholders if a successful outcome is to be achieved. Communication is also inexorably linked to effective leadership. Leadership qualities have a special significance in project-based sectors, as the project manager is a single integrative source of responsibility (Partington, 2003). The leader must effectively communicate the vision of what the organisation is trying to achieve (Thompson and McHugh, 2002: 260).

Construction project managers must apply appropriate leadership qualities to complex and multifarious situations throughout a project's lifetime and according to the level at which they are managing. According to Turner (1998: 16) these levels are defined as:

- *The Integrative level* – at this level the purpose and objectives of the project are defined, along with the risks and assumptions inherent with the activity. Resource requirements and constraints are determined.
- *The Strategic or Administrative level* – intermediate goals and/or milestones for achieving the project objects are defined and work packages are scheduled. This provides a framework for the day-to-day management of the project.
- *The Tactical or Operational level* – the activities required to achieve each milestone are defined and individual responsibilities for tasks are defined.

Each of these management levels defines a different range of communications challenges and competencies for the construction project manager. At the integrative level, the focus must be on communicating the *vision* for the project and ensuring that the resources allocated are adequate for meeting the espoused objectives. At the strategic level, the emphasis is on communicating achievable *targets* and goals and defining the boundaries around which the project will be broken down. Tactical management requires the individual *tasks* to be conveyed to the project participants in such a way as their individual contribution remains congruent with the overall project objectives and implementation plan defined at the Integrative and Strategic levels. The ability of the construction project manager to switch between these different communication functions defines it as one of the most demanding roles within the industry.

Summary

This chapter has explored the nature of project-based construction working and its reliance on effective communication. It has highlighted the inexorable relationship between effective communication and the performance of construction projects, highlighting the factors that can impede open communications within the construction project environment. Given that construction is such a fragmented, dynamic and disparate sector, the challenges of communicating effectively are manifestly greater than in most other production environments. Contractually driven relationships, conflict and a lack of mutual respect and trust, all combine to stymie open communication and render the role of the project manager extremely demanding and problematic. Nevertheless, addressing communication in the industry can be seen as a principal enabler for improving the industry in the future. In subsequent chapters, the barriers and challenges discussed here will be further elaborated in relation to their impact on interpersonal relationships, workgroups and teams, organisations and corporate communications. Techniques for overcoming them will be explored and approaches identified for helping the industry's project managers to develop high performance teams founded on an open communications climate and reciprocal knowledge sharing.

Critical discussion questions

1   Consider the issuing of a change instruction to amend the specification of a structural ground floor slab for a simple factory building. Identify the parties to which the change order must be communicated and count the number of interfaces through which the change order is likely to pass. Next, evaluate and discuss the difficulties inherent in ensuring the outcome of the order meets the requirements of the client.

2 Discuss the key communication challenges facing a project manager overseeing the construction of an office refurbishment project on a confined site in London. Your answer should consider the particular communication issues inherent with this type of project constructed in this type of location.

Case study: the importance of effective communication: the design and construction of the New Scottish Parliament in Edinburgh 1997–2004

Introduction

This case study pathology of the Scottish Parliament project examines how poor project communication can undermine the success of a high profile public sector project. It demonstrates the crucial importance of communication in supporting the successful outcome of large and complex projects from both a structural and cultural perspective.

The New Scottish Parliament Project in Edinburgh was completed in 2004 after a series of well-publicised set backs. These included the untimely deaths of the key client (Scotland's First Minister Donald Dewar) and the architect (Enric Miralles) and substantial cost and programme overruns. The project was the most prestigious construction project in Scotland's recent history and was a landmark for the process of Scottish devolution. Thus, the project was high profile in every sense and, even after completion, sections of the media continued to criticise the project for its apparent failures. Concerns over performance on this project have largely concentrated on the procurement and management of the scheme. It has been the subject of several enquiry reports that have given the public access to various aspects of the project. Of these enquiries, the 'Fraser' report (Fraser, 2004), offers the deepest insights into the workings of a major public construction project. However, each report has provided evidence of communication problems that have beset the project since its inception. These offer the opportunity to explore the anatomy of this major public construction project and the possibility of understanding the ways in which communication problems can manifest themselves within the industry.

Background to the project

In May 1997, the UK General Election returned a Labour Government which was committed to holding a referendum on Devolved Government in Scotland. By September that year, the referendum results indicated that almost 75 per cent of those voting agreed that there should be a Scottish Parliament. This result made it necessary to identify a permanent home for the forthcoming Parliament. The devolution White Paper estimated that the cost of constructing a new building for the Parliament would be between £10 million and £40 million. This estimate was made prior to the identification of a location or a design. Several locations in Edinburgh were considered for the new Parliament site, and Holyrood was announced by the Secretary of State in January 1998. The Holyrood site is located at the foot of Edinburgh's historic Royal Mile next to the Royal Palace of Holyrood House and Holyrood Park. The site had a long history as part of the medieval Old Town. In early 1998, it was occupied by a brewery.

The key players

The key players involved in the project are listed here. The project organisational structure is shown in Figure 2.2.

- *The Client* – Until the 1 June 1999, the client was the Secretary of State for Scotland. On 1 June 1999, the client responsibility transferred to the Scottish Parliament Corporate Body. From June 2000 the Holyrood Progress Group assisted the Corporate Body in their function as client.

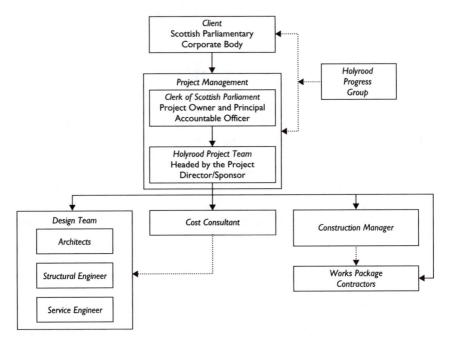

Figure 2.2 Holyrood Project Organization from June 2000.

Source: Black (2004).

Note
This shows the position from 20 June 2000 when the Holyrood Progress Group was established. Before 1 June 1999 the Client was the Secretary of State.

- *Project Management* – The project 'owner' and the project 'team' together constituted the project's management team. The Clerk and Principal Accountable Officer of the Scottish Parliament were designated the project owner, the most senior official responsible since June 1999 for the successful delivery of the project to the client. The Clerk is also responsible for ensuring that the parliament and the Corporate Body receive sufficient, informed and independent advice about the project. The project team included civil servants and private sector appointees on secondment to the team. They were responsible through the project director to the project owner for managing and delivering the project.
- *The Design Team* – EMBT/RMJM Limited had lead responsibility for the design of the new building. They were appointed in July 1998 after a competition. The other members of the design team were the structural and M&E engineering services consultants. They were chosen by the architects and approved and appointed by project management in 1998.
- *The Construction Manager* – The construction manager was appointed in January 1999 after a competition. They oversaw and managed the design and construction process on behalf of the client who remains the employer for all contracts. Under the construction management procurement route, they provided a service for the client and so were independent of the contractors responsible for actually constructing the building.
- *The Cost Consultant* – The cost consultant advised and acted for project management and on cost matters. Project Management appointed them in April 1998 after a competition.

Project communications

This analysis only discusses the key communication problems that took place during the life of the project. It does not discuss other aspects of the project, for which the full reports on the project should be consulted. The discussion that follows should provide several pointers on approaches to avoid when managing a major public construction project. The discussion is divided into sections and covers issues listed here. Several of these issues concern aspects of missing and incomplete information, gate keeping, 'noise' in communication, and the use of information as a control mechanism to influence others.

Selection of the Holyrood site

In June 1997 four possible options for the Parliamentary complex were considered. They covered a range of possibilities, from utilising the Old

Royal High School or St Andrew's House, to a site near Victoria Quay or a new-build on a green field site. The Holyrood Brewery site was on a much earlier short list but was discounted because of its constrained nature. As feasibility studies continued on the three principal sites, the Holyrood site was readmitted as a potential candidate. This set the wheels in motion for the eventual selection of the Holyrood site in January 1998.

Considering that the First Minister Donald Dewar was driving the process in a 'fast-track' manner, little time was available for a feasibility study of the Holyrood site and a local architectural practice was given only a week to conduct the study. Fraser found that the feasibility reports for the other potential sites were undertaken by different architectural practices, and as a result, it was difficult to compare the sites effectively. During the subsequent decision-making process, Donald Dewar was presented with indicative construction costs for a conventional building located on the sites under consideration. The Fraser report observes that the ministers should have questioned these indicative build estimates given the unconventional design solution that was likely to emerge from the competition they had in mind. Clearly, the final decision was influenced by the need for a budget which was sufficiently realistic, yet sufficiently low to allow for its political acceptability (see also Davidson and Huot, 1989; Connaughton, 1993; Mott MacDonald 2002).

Professional appointments, prequalification and tendering issues

In January 1998 the design competition was formerly announced. As well as criticising the speed in which this process was run, Fraser found that the evaluation procedure and scoring of the 70 returned prequalification questionnaires (PQQs) was less than objective with no methodical approach to short-listing. The process did, however, result in a shortlist of twelve possible contenders; later reduced to five including Spanish practice Enric Miralles y Moya (EMBT) who later joined the Scottish practice RMJM Ltd as a joint venture. On the 22 June 1998, EMBT/RMJM Ltd were selected as the architects. The panel members who were scoring the five competitive proposals were said to have been impressed by the joint venture team and in particular Miralles' sensitivity to the location of the Holyrood site and its proximity to Holyrood palace and the Royal Park. Fraser notes that his vision of a parliament building which 'sits in the land because it belongs in the land' was persuasive and may have struck a chord with Donald Dewar and his colleagues. However, the legal arrangements surrounding the incorporation of the joint venture company were described by Fraser as 'sloppy, unprofessional and fraught with danger'. Apparently, the joint venture company was only formerly ratified on 9 September which suggests that the initial Minute of Agreement dated 7 August with the Secretary of State for Scotland and signed by Enric Miralles purportedly on behalf of EMBT/RMJM Ltd was without contractual validity according to the Fraser Report.

Appointment of the construction manager

The appointment of the construction manager was also characterised by some irregular decision-making. From the 29 companies who responded to the OJEC Notice that appeared in the Official Journal on 12 August 1998, 15 returned the Pre-Qualification questionnaires by the closing date, and these were subsequently reduced to 6 who would attend for interview. Fraser stated that he was unhappy with the process undertaken by the project sponsor's decision to select the construction manager for the project. The decision was apparently primarily based on good reports of the work that they had done in another project in Edinburgh. The Crown Office and Procurator Fiscal service concluded that 'in the absence of any evidence to support an inference of criminality, there is no basis for a police investigation into whether or not EU or domestic procurement law had been breached' (*Contract Journal*, 2004a).

Appointment of key sub-contractors

The supplier for the cladding and windows package was Flour City Architectural Metals (UK) Ltd, a UK registered member of a group of companies owned by Flour City International Inc. registered in the United States. The company had no track record as a supplier in its own right and its capability to perform the contract was based on the reputation and record of its overseas holding company. The contract started in December 2000 but by August 2001 there were doubts about the ability of Flour City to perform the works. By October that year, Flour City (UK) underwent compulsory winding-up for insolvency. An Audit Scotland (2002) report into the impact of this revealed that the termination of Flour City's contract and subsequent re-tendering of the work would cost the client around £3.9 million. The report revealed that the appraisal of Flour City's financial standing was not sufficiently thorough and that even before the execution of the full trade contract, warning signs were evident. Flour City had requested advance payments for works, had not been paying its own suppliers and was aggressive in their pursuit of early payment of valuations. The report concluded that the award of the contract to Flour City was not improperly made, but that there were deficiencies in the selection, award and management procedures for the contract.

Defining roles and responsibilities

The first indication of a project wide document that would indicate roles and responsibilities was a revised copy of the original project brief, prepared by the project manager, and issued around May 1998. The document contained various appendices including sections on the roles and activities of the Project Manager, and Design Team responsibilities. The use of a project procedures manual may have resolved many of the communication problems that were to emerge in the Holyrood project. Fraser found evidence of a draft project execution plan and in the Auditor General's first report on the project (see Black, 2000) he refers to a draft copy of a project execution plan that was prepared in 1999. It was issued to all parties for comment and at the time of publishing his report it was agreed upon or finalised. In his 2004 report, Black reiterates the importance of this document 'a vital control procedure in any major construction project'. Furthermore, he praised the draft copy of the document which was comprehensive, setting out policies, strategies, lines of communication, key interfaces and responsibilities. However, despite significant changes in the project organisation since 2000, Black found that the document had never been updated which eventually led to an unclear definition of roles that beset the project.

Fraser's investigation into the use of the draft project execution plan revealed that the responsibility for its development belonged to the Project Sponsor and that this was in line with HM Treasury (1997) guidance. This also follows the guidance set out in an earlier report on the use of the Construction Management procurement form. Published by the Centre for Strategic Studies in Construction (CSSC, 1991) it refers to the need for a 'project control plan' developed from the client's brief and as such its formulation is the responsibility of the client. As the project develops responsibility, it is passed on to the construction manager to administer and update in conjunction with the designer and specialist contractors as and when they become involved in the project. This is where the Scottish Parliament project appeared to depart from this guidance. Fraser found that the draft project execution plan was to complement that issued by the construction manager for its own management and quality purposes. However, it would appear that the document was not regularly updated, which led to an unclear definition of roles during the project.

Anecdotal evidence supports Fraser's observations. One project participant, Lewis (2004) (a partner in the project engineers) was particularly critical about a lack of leadership:

> in any project there has to be a leader who coordinates and takes the project forward in a controlled fashion. It can be the architect, the project manager or the lead consultant. In this project there was no one, in my opinion, specifically to do that role.

Moreover he was critical of the communication channels on the project:

> We agreed through the spring of 1999 to visit Barcelona frequently for design sessions with Enric. These were productive but frequently invalidated by the following week, when we would be invited to view a new model of some part of the building.

This opinion was also borne out by the Auditor General (Black, 2004), who found that the 'organisation of the Holyrood project did not provide the necessary clear direction and leadership'. Moreover, he found that despite the project organisational chart (see Figure 2.2) suggesting a clear line of responsibility, control, communication and accountability, the actual position was much less clear.

The failure of the brief as a communication document

The communication of the client's requirements and the subsequent development of a project brief took place in late 1997 with the newly appointed Project Manager making fact-finding visits to parliament buildings in London, Dublin, and Oslo. The project brief was to be based on the Dresden Parliament building. Following consultation with ministers the brief evolved to a point where it would be submitted to the OJEC. However, the client (the Scottish Office) did not issue separate strategic and detailed Project Briefs for the new Parliament building. Fraser comments on the Auditor General's (2000) conclusions that the original version of the brief did not address the potential for conflict between the various dimensions of area, cost, time and quality. Moreover, he agreed that the brief did not recognise that client needs might evolve. This was evident when the client requested a radical change in gross area that went from 18,550 m^2 in 1998 to 30,593 m^2 in 2000. The inability to freeze the project brief may be a factor that is inevitable in a project procured under the construction management route.

Fraser's analysis of the briefing process highlights the inability to control a key communication activity that would eventually characterise the whole project. The failure to prioritise cost, quality and time issues led to a project where the evolving designs could not be properly costed, and this was evident in the number of packages where the original provisional sums were substantially lower than the tender figures submitted. It was therefore difficult to make clear cut evaluations from a briefing document that was not robust enough to control the design evolution. As Fraser observed:

> It suggests to me that over that crucial period in the development of the Project, sight was lost of the terms of the Brief. If that is correct, much of the extensive design development over that period was not taking place against the background of the clearly formulated set of client or user requirements, which the Brief should have contained.

Project relationships

The appointment of Enric Miralles was not without risks, and members of the review panel had concerns about whether he would be 'controllable'. During the early stages of the project the PM concluded that the Architect had made little progress since being appointed in July 1998 and estimated that by his programme they were already four weeks behind schedule in supplying drawings. This led the director of the joint venture company, who was based at the Edinburgh offices of RMJM, to express concerns about the difficulties they had been having communicating with the Barcelona office during the holiday period, the different 'operational characteristics' of the Barcelona and Edinburgh offices and the amount of abortive work that RMJM had undertaken. This was to set the tone for further communication difficulties that were a result of cultural misunderstandings between Edinburgh and Barcelona. As Fraser pointed out:

> The two practices had very different cultures and ways of working and found it difficult to adopt a cohesive approach to design issues or resolving problems whilst working in separate locations and communicating mainly via fax. With Enric Miralles insisting on being personally involved in all design issues during these formative stages, there was inevitable delay and disruption caused by his geographical detachment. Although RMJM took steps to better integrate the practices, this was only partially successful. Communication issues have been evident throughout the life of the joint venture company.

The lack of coordinated communication between the Edinburgh and Barcelona offices became most evident when in November 1998, RMJM and EMBT both submitted separate solutions to the Stage C difficulties to the Project Team; a course of events that exposed tensions between the two arms of the joint venture company.

Gate keeping

Two aspects of the project in particular provide evidence of 'gate keeping'. First the Fraser inquiry revealed many new insights into the workings of the Client body, and no more so than the revelation regarding 'edited' minutes. Apparently, in June 1999 the SPCB decided that, for reasons of commercial confidentiality, full minutes of its meetings would not be released to members of the Scottish Parliament or the public. Instead, an edited summary would be made available which would be as informative as possible. Fraser found that this practice was difficult to reconcile with the principles of openness and transparency. This practice was stopped in June 2000 following legal advice. A second issue concerned an apparent inability to communicate progress in the works. Fraser concluded that the construction manager had been reporting to the client with a 'degree of optimism' that was not justified and that the programmes reflected the political imperative for early completion.

Conclusions and lessons learned

The Scottish Parliament project provides many lessons in the importance of effective communication to the efficient management and delivery of large and complex construction projects. Whilst some of the events which combined to lead to the cost and time overruns for which the project is now renowned, a robust and effective communication structure, with clearly defined responsibilities for those involved, would have rendered such problems more manageable. In contrast, the Scottish Parliament project shows evidence of: informal decision-making (e.g. the process surrounding the selection of the site); informality (e.g. fuzzy decision-making surrounding the appointment of some of the project participants); unclear lines of responsibility and communication (e.g. in the interaction between the project sponsor, managers and other parties involved); a lack of coordination between the parties (e.g. between the members of the design team). Clearly defined responsibilities are clearly a cornerstone of successfully delivering such a complex project. The project client, client's management team, site appraisal team and all of the other players should have been clearly defined from the outset with clear reporting responsibilities and lines of communication defined for each. Ensuring consistency in the way in which the parties kept each other informed of their aspects of the work would have reduced confusion and would have ensured accountability for the problems that did arise. Without such structures, problems were left unaddressed for such a time until they could not be resolved satisfactorily. An important related factor was the need for effective project leadership from both the client and the other parties involved. Without a single point of responsibility, it was clear that no single individual or organisation was driving and coordinating the project and managing its outcomes. This gave rise to gate keeping and eventually to public confidence in the delivery team being undermined. Despite the high quality of professionals, managers and civil servants that were involved with the project, the Scottish Parliament example shows that, unless they operate within a well defined communication structure, they will not be able to manage such a complex undertaking successfully. This book will examine the ways in which the types of communications breakdowns revealed within this case study can be overcome.

Chapter 3

Theoretical perspectives on construction communication

The process of communication is supported by, and in some respects defined in terms of, an extant body of theory. This body of knowledge, which has evolved throughout the latter half of the twentieth century, has shaped understanding of the way in which people communicate, both formally and informally. It is fundamental to the understanding of human interaction and the ways in which information is sent, received and interpreted in the modern business environment. Theoretical perspectives on communication have important practical implications as they can inform decisions on how to communicate most efficiently and effectively (Skyttner, 1998). Accordingly, this chapter briefly expounds the theory of communication between individuals, groups and organisations. The perspectives explored provide an important foundation for exploring ways of improving communication within construction projects and organisations in subsequent chapters.

The development and value of communication theory

Although its origin dates back many hundreds of years, it is the work of the leading theorists of the twentieth century that have defined what we now understand as communication theory. Rather than developing as a coherent body of knowledge, it comprises a set of fairly disparate areas and sub-fields, many of which are rooted in the mathematical theories of Shannon and Weaver (1949), as well as the social and psychological perspectives of the late twentieth century. The study of communication in organisations traditionally examines the processing and flow of information through channels and networks (Thompson and McHugh, 2002: 260). In the 1970s, communications researchers began to explore the idea that organisations are processes and not entities, and are therefore defined, created and developed by communication itself (Hawes, 1990). This view sees communication as the essence of organisation and as such, fundamental to understanding how and why businesses succeed or fail.

An alternative view advanced by those who dismiss communications theory is that it is irrelevant or too far removed from the actual practices that are found in real-life situations. Such a perspective sees the communication process as a highly complex and subjective phenomena that cannot be understood through reference to simplified theories and models. This view arguably misinterprets the purpose and value of theory, however, which is to facilitate understanding of complex situations. Theories are used to explain and describe phenomena, and as such are the logical combinations of thoughts and ideas of observers. Using theories helps people to predict and control the phenomena. As such, theory and practice can be seen as mutually supportive. It is this perspective that is adopted in this book where theory is used to help make sense of the complex interactions that define communication within the sector.

In trying to gain an understanding of the different types of communication theory, it is useful to classify the levels of communication within which people are involved. One such taxonomy is provided by Kreps (1989), who divides communication into four levels which involve progressively more people within the communication process:

- *intrapersonal communication* (internal processes that enable individuals to process and interpret information);
- *interpersonal communication* (between two people in order that a person can establish and maintain relationships);
- *small-group communication* (more than two people communicating to allow them to co-ordinate activities); and
- *multi-group communication* (different work-groups communicating to each other).

In their book on construction communication, Emmitt and Gorse (2003: 45) also add the concept of *mass communication* (messages sent through the media or to large audiences) to this model. All of these levels can be seen to operate within the construction project environment. For example: intrapersonal communication in the way in which an architect interprets a client's brief and translates this into a building design; interpersonal communication in the direct interactions between a manager and their subordinates; small-group communication between members of the construction management and client teams; multi-group communication between a contractor's team and their sub-contract package contractors; and mass communication between a contractor's head office and their site-based staff. It is useful to view communication within these contexts in order to facilitate our understanding of it, and such an approach is adopted within this book. However, this does not infer that these communication types exist in isolation as all are likely to occur simultaneously within most real-world settings. This is one reason as to why understanding where a communication failure has occurred is often problematic (see case study in Chapter 2).

Theoretical models of communication

In Chapter 1, communication was defined as transmitting messages from one person and the receiving (and successful understanding) of those messages by another (Torrington and Hall, 1998: 112). Such a simplified definition, however, belies the complex mêlée of factors that shape and impact on the way in which messages are transferred from the transmitter to the receiver. This process and the variables that mediate them is extremely complex and thus, difficult to comprehend. Accordingly, many communication writers and theorists have drawn upon the simple analogy between the human communication process and the electronic telecommunications process where information is sent from a transmitter to receiver whilst being mediated by noise or distortion (e.g. Shannon and Weaver, 1949; Torrington *et al.*, 1995; Emmitt and Gorse, 2003). This simple analogy underpins the various theories of communication presented here.

Communication as a linear process

The most common application of communication theory is in understanding the transfer of information in the form of acoustic or visual messages, the function of which is to convey meaning. This differs from communication between machines, where the function is to control or process information (Skyttner, 1998). Shannon and Weaver (1949) developed a simple model of communication by electronic systems which, nevertheless has relevance to the ways in which people communicate, interpret and disseminate information (see Figure 3.1). This simple model comprises a transmitter relaying information as a signal to a receiver, the efficacy of which is affected by noise, which distorts the clarity of the message to the receiver, who subsequently decodes the message. In this model, every transmission has to be encoded in order for it to be communicated because otherwise the message could not propagate (Skyttner, 1998). Indeed, a message that has not been encoded would effectively have had to have been telepathically communicated!

'Noise' is important part of this process because it can impair the transmission of the message from the sender to the receiver. This can account

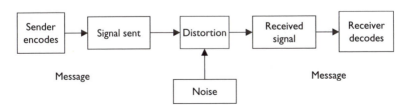

Figure 3.1 A model of the communication process.

Source: Adapted from Shannon and Weaver (1949) & Emmitt and Gorse (2003).

for any type of distortion or distraction that can affect the quality of transmission between parties. A well-cited example of this is the 'send three and fourpence, am going to a dance' message from the First World War. This message relates to a communication which was verbally communicated in relays from the front line to headquarters, but the message received was not that which was sent. Along the relay, a number of cumulative errors had occurred as individuals sought to make sense of (decode) what they thought they had heard and then pass it on (encode) to the next person. The message that was initially encoded was 'Send reinforcements. Am going to advance'. This is considerably different to what was eventually decoded at headquarters. Applying this simple model to the real world of communicating within the construction industry, the issue of 'noise' is particularly relevant, as there are many factors which can distort the clarity of a message relayed between individuals and teams. These include the physical distance of the sender from the receiver. Messages are often relayed between offices or from offices to sites over some considerable distance. Another source of noise stems from the probability of misunderstanding the type of technical information flowing between participants in the construction process. The technical language used and the various frames of reference that exist within the (often fragmented) construction process effectively increases the likelihood that meaning will be misinterpreted by one or more of the actors involved. In addition, the actual physical noise (plant and machinery, etc.) associated with construction activity can prevent the clear transfer of verbal messages from one person to another. Together, this presents a particularly problematic communication context as was discussed in Chapter 2.

It is fairly obvious that Shannon and Weaver's simple model masks a far more complex set of parameters that combine to shape the way in which information is transferred and interpreted and has been widely criticised as a result. For example, it ignores the fact that all communication is potentially a two-way process, with the receiver (if there is one) providing feedback to the sender on whether the message is understood. Furthermore, it also ignores the physical and social context in which people work. An individual's interpretation of this information is wholly reliant upon their personal perspective or perception and their individual frame of reference. Certain individuals will choose to ignore information that they do not consider relevant to their own particular circumstances. Thus, although it provides a convenient starting point for understanding the communication process, it fails to satisfactorily convey the process of human interaction.

The effects of feedback on the communication process

An example of a more sophisticated version of this model is provided by Baguley (1994), which portrays communication as a two-way rather than

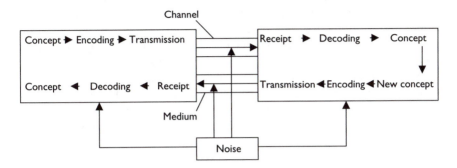

Figure 3.2 The communication process.

Source: Philip Baguley (1994) *Effective Communication for Modern Businesses*, McGraw-Hill, London, p. 9.

as a linear process. This is depicted in Figure 3.2. This model includes reference to the communication medium and channel along which it passes. Viewing communication as a process in this way presents it as a more dynamic or iterative concept where the transmitter is continually receiving feedback (albeit still moderated by noise). The *medium* refers to the means of communication used (such as spoken, written, graphic etc.) and the *channel* is the conduit through which the message passes (such as the telephone, a meeting, a letter etc.). The feedback is more than a simple acknowledgement of the message, as it should contain as much information as the original message.

In this model, the relevance of the *manner* in which information is transferred is taken into account, as are the internal processes of the transmitter and receiver and the fact that noise can be present *within* these processes as well as in the actual transfer through the communication channel. For example, the effectiveness of communicating a design detail will depend upon how well it is conveyed, which will usually be through graphical images such as CAD drawings. Clearly, describing such a detail using words alone would be more difficult with there being less chance of the design detail being understood properly. Thus, this model offers a view of communication that is more representative of the realities of the process as it takes into account the way in which communication takes place between the transmitter and receiver.

The impact of cultural context on the communication process

Although Baguley's model is more powerful in reflecting the continual and dynamic nature of communication between people, it fails to consider all of the external environmental factors (other than noise) that can impact upon the efficacy of the communication process. This 'closed system' view of communication therefore ignores the context within which communication takes place. A more advanced model is arguably provided by Thompson and McHugh (2002: 261). This is based on Fisher's (1993) model of communication in organisations and adds the dimension of *context* to the communication process in the form of the structures, cultures, group task characteristics and information from the environment that combine to impede or promote the efficacy of the communication process. Collectively these present a wide range of what are termed mediating variables. Indeed, as Thompson and McHugh themselves remark, given the range of factors affecting the process it is surprising that any successful communication occurs at all! Nonetheless, the model accounts for the shared assumptions, beliefs and rituals embodied within organisational culture (see Schein, 1985), and so presents an 'open systems' view of the communication process. It has close links to the perspectives on organisational communication adopted by HRM theorists (discussed here). Figure 3.3 develops Thompson and McHugh's model for the construction industry.

Despite the increasing sophistication of these models in representing the communication process, they all have a central weakness in that they view communication as a *sequential* rather than a simultaneous or concurrent process (Thompson and McHugh, 2002: 261). This effectively ignores the interpersonal dynamics of communication which characterises most forms of human interaction. As such, it must be augmented by the social

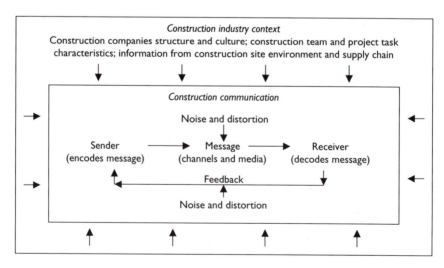

Figure 3.3 Construction industry context and communication process.

Source: Based on: Thompson and McHugh (2002); cf. Fisher (1993).

and organisational contexts which give rise to the meanings that allow communications to take place. These include the expectations of the actors (which will in turn be grounded in their individual backgrounds, needs, perceptions and goals). Thus, although they are useful for conceptualising communication as a process, this masks the complex interplay of signs, meanings and symbols (semiotics) which are often more important than the explicit communication itself (see Fiske, 1990). Whereas the process perspective sees communication as the transmission of messages through which one person seeks to influence another (and hence focuses on how transmitters and receivers encode and decode messages), the semiotic method sees communication as the development and exchange of meaning. Thus, unsuccessful communication may not be a product of process failure, but of misunderstandings grounded in cultural differences with regard to meaning.

The importance of culture, especially in terms of interpreting non-verbal behaviours, cannot be overstated. For example, what is a complimentary or reverential sign in one country may be a direct insult or curse in another. Moreover, different subcultures exist within different societies and even organisations which can lead to radically different interpretations being placed on identical gestures. Hall (1989) distinguishes between high- and low-context cultures. High context cultures (such as China and Japan) place more importance on signs and status of relationships than what is actually said, although agreements tend to be made very informally. Low-context cultures such as the USA pay more attention to verbal messages, although agreements are rarely made until a legal document is signed. Thus, the way in which business is done will differ depending upon the national cultural context in which it is undertaken.

Applying communication models to construction

The communication perspectives outlined earlier can be used to diagnose reasons as to why actors within construction projects often fail to communicate effectively and to determine new approaches towards improving communications performance in the future. The effectiveness of communication in all of the models discussed earlier essentially rests upon four factors:

- The effectiveness by which information in *encoded* and then *transmitted* through communication systems, channels and networks.
- The appropriateness of the communication *mediums and channels* used.
- How those receiving the communication *decode, interpret and act* upon it.
- The abilities of those commentating to *minimise 'noise'* which could impede the process.

In reality, these four factors are likely to be accounted for simultaneously by the actors involved through the ways they have developed to communicate effectively within their workplace environment. It is worth exploring each aspect in turn however, in order that the principles of effective communication can be established.

Encoding and transmission

The effective encoding of information has been a major problem in the construction industry for many years. For example, drawings are often issued incomplete or presented in such a way as to contradict other project documents. Many of these problems originate in the plethora of different occupational cultures which characterise the industry, each having different educations, languages and customs which others find difficult to understand (Moore and Dainty, 2000). A good example of this has been the problem of construction contracts which, to the uninitiated, are very difficult to understand due to their complex structure and legalistic language. Historically, this been a major cause of conflict in the construction industry (Latham, 1994). The divergent occupational cultures which give rise to such problems can be traced back to organisational changes in the construction industry between the middle ages and the nineteenth century, which led to the demise of individual artisans and to the domination of influential capitalist contractors supported by waged labour (Musgrave, 1994). These changes led to the development of the professions and a hierarchical structure of social superiority (Hindle and Muller, 1996). Over time, professional institutions have developed vested interests in maintaining their social status by developing specific methods of communication which are reinforced by exerting control over the educational system and standard forms of construction contracts. Thus, encoding within the industry takes on many different forms to reflect the vested interests of the institutions of which it comprises.

Communication media and channels in construction

Having encoded a message, it then has to be communicated through effective communication channels. This is particularly challenging in construction because of the number of boundaries across which communication must pass. An increasing use of project partnerships, strategic alliances and efforts to improve client satisfaction are extending the need to communicate beyond organisational boundaries and to embrace project delivery structures and their supply chains. This is because employees within construction companies do not operate in isolation, but must liaise and interact with the suppliers and customers who surround them. For example, contracting organisations outsource the majority of their work and buy in specialist design services where required. These service providers may be outsiders to the organisation in general, but they form integral members of the project delivery team. Partnerships between supply chain partners cannot work unless effective communications channels exist within and between the organisations involved. This presents a real challenge for companies at the centre of project delivery (usually main contractors) who must maintain a range of complex channels with different types of organisations.

The medium through which information is transacted will also have an important bearing upon the efficacy of the empowerment process. In construction, a useful classification is to see media as either *one-way* or *two-way*. One-way media facilitate a linear and unidirectional flow of information from sender to receiver with no opportunities for feedback. This might include letters, drawings, emails or faxes between parties in the supply chain. In contrast, two-way media facilitate a reciprocal and multi-directional flow of information from sender to receiver with numerous opportunities for feedback. This might include meetings, telephone calls or conferences. If the aim is to secure trust, mutuality and a spirit of cooperation amongst the members of the project delivery team, then two-way media are preferable. One-way media have the potential to undermine the environment necessary for involvement and empowerment within the workplace. However, they are also cheaper and more rapid ways to convey information and this is why they are increasingly relied upon in construction projects, even where meetings might have been more appropriate. Thus, it is important to note that the media used to communicate in construction is not always the most appropriate in terms of the transmitting and understanding of meaning.

Interpretation and action

It is essential for any construction project manager to be aware of natural human tendencies with regard to the interpersonal communication process. Armstrong (2001) defines the key barriers to effective communication in this regard as:

- people hearing what they want to hear and ignoring conflicting information;
- people having perceptions (and preconceptions) about the communicator;
- influence of the group dynamics (people are more likely to listen to the opinions of their own group than those of outsiders);
- words meaning different things to different people (terminology and language barriers);
- non-verbal communication issues (misinterpretation of body language and other non-verbal signals);
- emotions (which can colour the reception of the message);

- noise (in both the literal sense and other distractions to the process of listening to assimilating information); and
- size of the organisation (as a general principle, the larger it is, the more difficult effective communication is to achieve).

Taking an open approach to communicating with people can enhance the likelihood that people will react in the desired manner. This means being fair, equitable, honest and clear about information transfer and not manipulating it in favour of one or other party. In reality, those choosing to transmit the information must decide the degree to which people need to know it. In this regard, information is 'power', particularly when it deals with commercially sensitive matters. There is evidence to show that this can lead to problems within the construction project environment. For example, Loosemore (2000) found that it is common for information to be manipulated by project members in their favour in relation to unexpected problems. The effect is that problem resolution can be prolonged and/or made more costly. Thus, in these closed communications systems, the real agenda behind the transfer of information is kept secret and hidden, and this erodes the trust-based culture which is necessary for effective teamwork.

Impact of 'noise'

'Noise' is important because it can impair the understanding of the message by the receiver, no matter how appropriate the channels and media that are selected. In construction, one source of noise problems concerns the physical environment of the workplace. Construction sites can be noisy and thus present many distractions which impair people's abilities to communicate by word of mouth. However, of even greater significance are the noise effects caused by interfaces in the message chain. This is because every person involved in the transfer of information will filter or amplify the message sent (Thomason, 1988: 409). This phenomenon is popularly known as 'Chinese Whispers' and can result in the message arriving at the intended recipient differing radically from that envisaged by its source as was discussed earlier in relation to the World War One example. In construction, formalised communication chains exist which can increase the likelihood of filtering amplification. For example, a client wishing to instruct a change to a building during the construction process will typically transmit this through their architect, who will in turn issue an instruction to the main contractor who will then inform the subcontractors and suppliers. This hierarchical process effectively increases the probability that noise will affect the clarity of the instruction sent. The industry has tried to circumvent this problem by developing legal protocols for issuing such change instructions, but these formalise what could otherwise have been a collaborative (and hence possibly more effective) process. For example, if the sub-contractor had been involved in the initial decision they may have found a more appropriate way to cope with the change, rather than the process being decided by the other parties involved.

Another manifestation of noise in construction concerns the emotional impact of those involved in the industry. Situations often arise when people do not communicate effectively because they allow their emotions to impede their communication. Reasons as to why this occurs in construction are speculative, but it could relate to a general lack of trust between the parties involved (see Latham, 1994), or could be grounded in the masculine culture of the industry brought about by male domination within the sector (see Gale, 1992). However, emotions are also fundamental for understanding why communication works effectively in construction. The working culture of the sector as manifested in people's interactions can lead to innovative problem-solving and creative thinking. Indeed, the importance of human interaction in construction is such that it may militate against the effectiveness of some type of distributed working (such as virtual teams – see Chapter 8) facilitated by advanced telecommunications systems in other sectors. This 'emotional impact noise' emphasises one of the major problems with communication theory, in that theory cannot account for all of the nuances of human behaviour in affecting the communication process. Nevertheless, it can be appreciated that an industry such as construction, which relies on human interaction for so many of its processes, is more likely to suffer from the effects of noise than those for which communication is less important.

Optimising communication in construction: the importance of the recipient

Given the barriers to effective communication outlined earlier, it stands to reason that there also a range of criteria which will form the optimal conditions necessary for effective communication. According to Skyttner (1998), the critical conditions that must be apparent for optimum communication performance are as follows:

- The information source must provide adequate and distinct information.
- The message must be correct and completely coded into a transmissible signal.
- The various kinds of noise must be taken into account if the signal is to be transmitted in a rapid and correct form.
- Received signals must be translated in a way that corresponds with the coding.
- The receiver must be able to convert the message into the desired response.

This set of guiding principles is based around ensuring clarity of information, the use of appropriate media and taking account of noise and distraction.

Of equal significance however, is the need to take account of the perspective of the receiver (or listener). With the exception of intra-personal communication, all human communication is inherently a two-way transaction between people, and effective listening is in fact the basis of all person(s)-to-person(s) communication (Sheldrick-Ross and Dewdney, 1998: 15). Indeed, In many respects it is the *recipient* who actually communicates, as unless somebody 'hears' there has been no communication, only noise. This emphasises both the two-way nature of communication (see Baguley, 1994: 7) and the crucial importance of the feedback element of the models presented earlier.

This principle is extremely important in the context of the construction sector because if the needs and perspective of the eventual recipient of the communication are not considered, the outcome is unlikely to accord with the needs of the sender. In Chapter 2 it was suggested that communication in construction will almost always transcend professional and functional boundaries. Hence, the sender of the information must consider the needs of the eventual recipient, even though they may speak a 'different language' or operate within a wholly different socio-cultural context. It is this failure to consider the appropriateness to the recipient of the media selected that undermines communication within the industry, as is explored further later. Nonetheless, attempting to put in place the right critical conditions for effective communication to take place is incumbent upon every manager within the industry.

Communication media

Communications can take many different forms, with each being appropriate for particular circumstances. The choice of which medium is the most appropriate will depend upon the nature of the information and recipient, and the outcomes desired from the communication. For example, a contractor requesting an extension of time would not be best communicating this through an informal exchange at a project social event, just as informing a project manager that s/he has been turned down for a promotion would not be best conveyed through a written memo alone. Although it is incumbent upon the transmitter to select the most appropriate communication medium (or combination of different media) for the information needing to be conveyed, messages are often more successfully conveyed if a variety of media are used (Torrington and Hall, 1998: 121).

In broad terms, communications media comprise the following generic types:

- *Speech/verbal* – oral communication between individuals or groups. Can be formal (such as in meetings and focus groups) or informal in nature, and face-to-face or via the telecommunications media.

- *Non-verbal* – present during some other forms of communication and implying meaning to what is being said. Can be deliberate or non-deliberate in nature.
- *Written* – usually official or formal in nature, written information provides a permanent record of the communication if desired.
- *Audiovisual* – graphical or audio-based media designed to convey a message more effectively.
- *Electronic* – increasingly popular method using innovations such as electronic mail to communicate rapidly between distributed individuals and groups.

Each of these media can be expressed formally or informally depending upon the circumstances and desire of the individual or group transmitting the message. Some of these media will be naturally combined during the communication process. For example, a spoken conversation between two people in close physical proximity will inevitably involve a degree of non-verbal behaviour in the form of body language. In other cases, media will be deliberately combined, such as where drawings are used to clarify a design change under consideration by two or more members of the design team. These types of communication media are discussed in more detail here.

Verbal communication

The spoken word could be described as the most direct form of communication. Most conversations involve an exchange of information. These tend to occur as a result of questions which come in many forms (open, closed, probing, hypothetical etc.). Conversations can be formal, informal, long or short and between individuals or groups. Conversations and discussions are often the most effective ways of ensuring that people feel involved or consulted in a process. They also allow immediate feedback to be collected by the transmitter and so have the potential to avoid some of the effects of noise that can militate against the effectiveness of other media. This does not infer that such interactions are without problems however, as will be discussed in Chapter 4.

Non-verbal communication

Direct communication is also affected through a range of non-verbal signals. Non-verbal cues convey the nuances of meaning and emotion that reinforce or contradict the verbal message in a given situation. Non-verbal messages are conveyed through 'behaviour' which comprises a range of direct and indirect cues (Sheldrick-Ross and Dewdney, 1998: 3). These may be conscious or unconscious and can either reinforce verbal cues or convey a completely different message depending upon how they are interpreted.

In many ways they can be more powerful than the verbal messages they are usually combined with. Individual non-verbal cues can come in the form of expressions, gestures and eye movements (particularly by the pupils). Some of these are voluntary and others involuntary, but all combine to relay more meaning to the verbal cues provided through conversation. Examples of non-verbal behaviours or body language include:

- *Eye movements* – the eyes are one of the most expressive organs of the body. Eye behaviour is known as occulesics and can include enlargement of the pupils and other movements around the eyes, eyelids or eyebrows. Maintaining eye contact is also seen as an important element of being perceived as trustworthy in many cultures. In other cultures, avoiding eye contact is regarded as an indication of respect.
- *Facial expressions* – contorting the face adds emotion to verbal cues which again differ depending upon the cultural context.
- *Posture* – the way in which someone stands or sits gives important cues as to the interest they hold in communicating with another person. For example, sitting slightly forward shows interest, where as slouching can convey disinterest or arrogance.
- *Body contact and touch* – known as tacesics, this is a powerful non-lingusitic way to display strength of feeling or feedback in relation to communication as well as being a strong communication mechanism in its own right.
- *Limb movements* – the movement of limbs during conversations, known as kinesics, can display a range of emotions in relation to the person being communicated with. For example, crossing the arms and legs can look defensive whereas, sitting with palms facing upwards gives the impression of openness or honesty.
- *Tone and pitch of voice* – this is known as paralanguage or vocalics and refers to the speech, pitch and loudness of the voice. The way in which something is said effectively influences the meaning of what is said and the interpretation of the receiver. Some languages rely heavily on the use of tone to imply meaning.
- *Distance* – different societies or cultures require different amounts of personal 'space' in order to engage comfortably in conversation. This phenomenon is known as proxemics, and differs wildly within different cultures and locations.
- *Timing of verbal exchanges* – known as chronemics, this refers to the way in which verbal exchanges are timed, their duration, urgency, waiting time and punctuality. Again, the prevailing business culture will dictate particular approaches which are deemed appropriate to exchanges between actors involved in a transaction.
- *Physical appearance* – the way that people look (e.g. their clothing and hair, grooming and cosmetics) provides messages as to an individuals

feelings or disposition. Again, an inappropriate appearance can convey disinterest or a lack of respect in certain business exchanges.

It can be appreciated that some of these non-verbal cues can be deliberately controlled and manipulated, whereas others represent unconscious 'body language' that people have little or no control over. As such, they can contradict or neutralise what is being said (consider someone speaking in a sarcastic manner for example). It is therefore important that managers gain an appreciation of non-verbal cues in order that they can put their messages across effectively and correctly interpret feedback from those with whom they have direct contact. Managers who do this effectively are more likely to interpret situations correctly and take action to mitigate any communication problems that may arise.

Clearly, important as they are, non-verbal cues within communication are not restricted to those instigated by the actors involved. The nature of the environment or situation can also have fundamental effects with regards to the quality of non-verbal communication. Richmond *et al.* (1991) define a range of non-verbal categories of communication from individual physical behaviour to the nature of the environment within which people are interacting. For example, the interplay of spatial arrangements such as music, lighting, colour and temperature can be used to send non-verbal messages. Thus, it can be seen that true understanding of any communication will depend upon the prevailing situation and environment as much as it will the effectiveness of the individual transmitting the information. Non-verbal cues are discussed in relation to construction projects in more detail in Chapter 4.

Written communication

Written communication is indirect in nature, but has a big advantage in that it provides a permanent record of the communication if desired. Many forms of written communication are governed by rules and/or protocols dictated by organisations or professional codes of conduct. For example, the way in which business letters and memos are sent is almost always determined by accepted standards which are adopted in the business world to facilitate understanding. In construction, standard forms are available for structuring many of the communications between the parties involved in projects. Written forms of communication allow the sender to carefully consider what they want to say and to convey this in a clear and meaningful way without the need for direct interaction with the recipient. Although this has disadvantages in the time it takes to receive feedback, it avoids the need to respond to questions arising from the communication as it is constructed. It also allows the sender to communicate with a number of different people at the same time using a medium that is readily understood and unlikely to be misinterpreted.

Audiovisual communication

As well as written forms of communication, graphical or audio-based media can help to convey a message more concisely and effectively. Consider the use of drawings in construction. Without these, designs would have to be described using the written word or through the construction of a series of models! Often, graphical media are used in combination with other media in order to convey information more clearly. For example, the use of graphs and pie charts allows the receiver to assimilate information far more rapidly than stating percentage figures. Hence, these are often used to report financial information in order to effectively summarise complex data. There are also advantages in conveying complex technical issues to parties who would otherwise struggle to interpret meaning within exchanges. For example, it would be easier for an architect to sketch a diagram showing a revised construction detail for a client than to trying to describe the various components using technical jargon, particularly if the client is unfamiliar with construction terminology and techniques.

Electronic communication

In recent years there has been a revolution in the development of ICT. Electronic forms of communication, such as email and web-based tools, are now commonplace in virtually all industries and sectors. The take-up of these types of technology has been particularly rapid because of the advantages that they present in speeding up the rate of transaction between parties and transferring information between parties. Moreover, they offer benefits in lowering the costs and environmental impacts associated with paper-based communication, particularly in replacing the necessity for face-to-face meetings. Innovations such as email and even video conferencing are now possible through hand-held communications devices which allow the user to communicate from remote locations and sites. Furthermore, the power of visualisation tools to facilitate 'virtual prototyping' and inform design decisions is now offering real possibilities for construction organisations to overcome many of the constraints outlined in Chapter 2 with regards to the uniqueness of construction products. These issues are discussed in greater depth in Chapter 8.

Choosing appropriate communication media

The factors underpinning the choice of which media to adopt will involve a degree of choice on the part of the transmitter. The first determining factor will be the objective of the communication. Consider a manager communicating to their staff. They must decide upon whether they are attempting to persuade, inform, question or instruct (Baguley, 1994: 11). They will then tailor their approach to suit the needs of the communication. This decision will be to some extent governed by the need for formality; informal messages are best communicated by informal means and vice versa. The second factor will be the *needs* of the communication. This will be determined by the nature of the message, the intended recipient, the channels available, the need to record the feedback of the message and the need for consistency. Baguley defines these factors as the 'how' of communications. Misinterpreting the objectives or needs of the communication is the most common reason as to why communications fail, as is explored below.

Barriers to effective communication

Although this chapter has presented a range of different theoretical perspectives on communication along with appropriate media and channels to ensure its success, inevitably communications will fail on some occasions. It is important to understand the theoretical barriers to effective communication in order that the tools and techniques explored later in this book can be appropriately applied to overcome them. Different authors have taken different perspectives on defining barriers to effective communication. Together, these present a range of challenges that must be overcome if communication is to be successful. Torrington and Hall (1998: 116) identify several barriers to effective communication including:

- *The individual's frame of reference* – people interpret meaning in communication in ways which are shaped by their own frame of reference. For example, cultural differences will shape the ways in which an employee responds to a new ruling on wearing health and safety equipment, as in some cultures this would be perceived as a doctrine that must be complied with, whereas in others, merely as advice to which the employee does not have to respond compliantly.
- *Stereotyping* – people have a tendency to stereotype others according to their socially constructed views. For example, rather than listening to what someone is saying, they will hear what they expect that person to say given their socio-economic background, profession or perceived disposition. Stereotyping can therefore lead to a course of action which does not accord with the needs of a particular situation.
- *Cognitive dissonance* – if someone receives information which conflicts with their established beliefs then they will often have difficulty in understanding or responding positively to it. Rather, they will disbelieve or challenge it as a way of dealing with the inherent discomfort of dissonance.
- *'Halo or horns' effect* – if someone is trusted by another individual, they may be predisposed to agree with what they say. Conversely, if someone is distrusted then what they say may be ignored or treated with caution. This is related to stereotyping, but is more closely related

to an individual's perception of another person or organisation rather than a distinct societal group.

- *Semantics/jargon* – this infers difficulties in transferring the meaning of information from one person to another. The meanings behind any piece of information are attributed to the receivers and are not embodied within the words themselves. If jargonised words are used then the receiver may have no idea as to the intended meaning of the communication.
- *Not paying attention* – being distracted by the other noise going on around a particular communication is a primary cause of communication difficulties, as is simply forgetting information soon after it has been communicated. A multitude of factors can impact on this phenomenon, such as occupational stress and burnout, which can adversely affect concentration and hence the performance of individual employees (Loosemore *et al.*, 2003).

Within organisations there is another set of variables which can act to impede successful communication. These are the structure of relationships and cultural and societal norms govern the power dynamics which can combine to affect the effectiveness of communication. According to Huczynski and Buchanan (2001: 183) there are five principal barriers to effective communication within the organisational setting which can be generalised as:

- *Power differences* – employees distort upward communication and believe that superiors have a limited understanding of subordinate's needs.
- *Gender differences* – men tend to talk more whereas women tend to listen and reflect more.
- *Physical surroundings* – issues such as room layout, noisy equipment and physical proximity affect communication effectiveness.
- *Language* – variations in language and dialect can affect communication.
- *Cultural diversity* – different cultures harbour dissimilar expectations as regards formal and informal expectations.

Huczynski and Buchanan's factors are grounded in the nature of human interaction and relations. People will interact with colleagues in ways which are influenced by their socialisation and cultural values. Given that communication is a two-way process, the role of the listener is just as important as that of the sender in terms of the overall impact of communication. Sheldrick-Ross and Dewdney (1998: 16) identify the typical barriers to effective communication from the listener's perspective:

- *Selective perception* – the listener only hearing those messages that support their model of the world and filtering out others.

- *Making assumptions* – the listener assuming what the sender means or feels rather than listening to what they say.
- *Giving unsolicited advice* – the listener giving unwanted advice or providing advice before listing carefully to the problem.
- *Being judgmental* – the listener being critical of another person's point of view in a way that distances them from the transmitter's point of view.
- *Acting defensively* – the listener defending a position rather than listening to the position of another person.
- *Failing to understand cultural differences* – subtle but significant differences in language or pronunciation often lead to miscommunication. For example, silences have distinct meanings and require specific etiquette in some cultures that must be respected for effective communication to take place.

These barriers infer that effective communication is reliant on effective listening skills as much as it is communication skills (Baguley, 1994: 33). Consider, for example, a project manager who is unwilling to listen to any feedback from his/her subordinates on how to manage or organise a process more effectively. Given that the project manager is unlikely to possess all of the skills and knowledge held by their staff team, they are highly unlikely to resolve the problem as effectively as if they had canvassed the opinions of their staff. Nevertheless, managers are often poor listeners and as such, the communication process can often be blighted by these factors.

Summary

This chapter has provided a basic introduction to the key points of relevant communication theory as a basis for more in-depth consideration of the practice of communication in subsequent chapters. It has presented a range of theoretical perspectives on the ways in which people communicate, together with the mediating factors which affect the quality and outcomes of communication. These theoretical perspectives offer convenient (if somewhat simplified) interpretations of communication that act as frames of reference to allow the analysis of the processes and semiotics involved. These theories have been related to the unique socio-cultural context of construction in terms of the ways that the industry's norms and processes govern encoding, the selection of media and channels, the interpretation of meaning and 'noise', and their affect on the efficacy of the process. The themes and issues covered within this chapter will be returned to later in this book as various levels of communication in construction are explored in-depth. They should also provide the reader with a 'toolbox' of techniques for analysing the communication problems emerging from the critical discussion questions provided at the end of each chapter.

Critical discussion questions

1   Identify the patterns of communication that are likely to determine the commercial success of the following construction projects and use the theoretical perspectives developed within this chapter to compare and contrast the communication flows and challenges likely to occur in each project:

    •   A simple single floor extension to a modern domestic dwelling.
    •   The construction of a new terminal building for a major airport.
    •   The refurbishment of a Grade II* listed swimming pool in a historic market town.

2   Consider the communication patterns for a project with which you are familiar. Discuss the impact that power differences, gender differences, physical surroundings, language and cultural diversity had on the efficacy of communication between the various project stakeholders.

Part II

From individuals to corporations

Communication types and techniques

This part explores the practice of communication in the industry, identifying how the constraints identified in Chapter 2 can be overcome through the effective use of communication principles, tools and techniques. The following chapters draw upon the theory expounded in Chapter 3, but within the practical context of communicating effectively within the construction industry. As with previous chapters, a particular emphasis is placed on exploring communication challenges from the perspective of the construction project manager.

Chapter 4

Interpersonal communication

One form of communication that takes place within construction project environments is that which occurs directly between individuals. The regular meetings that take place within and between members of design and construction teams will be supplemented by interactions amongst a multitude of project stakeholders such as clients, subcontractors, suppliers, local authority building inspectors, Health and Safety Executive inspectors, members of the public and others. Within these information exchanges, information will be transferred and conveyed via a range of formal and informal media in both deliberate and non-deliberate ways. As Richmond *et al.* (1991) note, verbal communication does not occur in a vacuum, but often occurs in combination with non-verbal communication processes. Accordingly, this chapter explores interaction between individuals, considering both the verbal and non-verbal cues that occur between people in construction and the ways in which communication can be made more effective. The majority of these communications will be take place between individuals using face-to-face meetings or through other media such as telephone, fax, e-mail and letter, all of which can be regarded as one-to-one or *interpersonal* communication. Although communications using postal and/or electronic means do not necessarily allow for issues such as body language and paralanguage to come into play (see Chapter 3), messages may still be misinterpreted or misunderstood if they are not conveyed effectively. Thus, the role of other media in interpersonal communications is also considered.

Defining interpersonal communication

Interpersonal communication generally refers to the process of communicating between two or more people. An interpersonal channel is one that involves a face-to-face exchange between a source and a receiver (Rogers and Agarwala-Rogers, 1976: 12). Whilst some writers refer to interpersonal communication in a group context, there are arguably many differences between the patterns of interaction manifested between individuals in comparison to groups of people (Emmitt and Gorse, 2003: 45). The way in which one person will attempt to communicate with another will depend upon how they assume that the recipient will interpret and respond to the information they wish to transmit to them. Effective interpersonal communication at all levels is, however, always crucial to the performance of construction projects as this chapter will reveal.

Although there can be any number of recipients of information within interpersonal communication, it is important to note that any communication between parties will also involve intrapersonal communication processes on behalf of both the source (sender) and recipient (receiver). Intrapersonal communication refers to the internal cognitive process of transmitting our thoughts and so may not always be evident in external emotions. As it is a cognitive process, theorists investigating this subject often refer to the work of psychologists and neurophysiologists to reinforce their definition. Whereas an in-depth discussion of intrapersonal thought processes is beyond the scope of this textbook (see Melvin (1979) for a more comprehensive/detailed exploration of psychology in construction management), it is important to note that all communication processes begin with intrapersonal communication before manifesting themselves in 'external' interpersonal communications.

The nature of effective interpersonal communication within construction

As is noted earlier, interpersonal communication can occur in many different ways using a variety of media. However, in an industry such as construction where any number of approaches could be relevant, it is important to first establish which approach yields the most effective results for the performance of its projects. Several researchers have examined the use and efficacy of different approaches within the construction project environment. Gorse *et al.* (1999) for example, investigated the dynamics of interpersonal communication methods and the behaviour between designers and contractors during the construction phase of projects. Their study sought to measure the perceptions of effectiveness of communication media, the most obvious examples being person-to-person interaction or the use of electronic technology as a conduit for transfer. Gorse *et al.* explored a range of media from informal approaches, such as face-to-face meetings to more formal methods such as letter, fax and email, their results showing that the former was perceived to be the most effective medium of communication within the industry. Gorse *et al.*'s findings are supported by Carlsson *et al.* (2001), who conducted communication research within the Swedish construction industry. Their findings also indicated a preference by construction personnel for face-to-face interaction. Table 4.1 shows the results from their case study projects. Projects A and D were procured by the design and build

Table 4.1 Methods used for communication (number of contacts, %)

Method	Project A	Project B	Project C	Project D	All projects
Meetings	51	74	61	34	57
Telephone	36	15	13	34	21
Fax	4	7	12	18	11
E-mail	0	2	7	6	5
Data files	1	1	3	3	2
Letters	2	1	3	1	2
Drawings	6	0	1	4	2
Total	100	100	100	100	100

Source: Carlsson et al. (2001).

route, whereas projects B and C were procured by the traditional route. It can be seen that the most used method of communication was face-to-face meetings (57 per cent), although their use was more prevalent in the projects procured under traditional contracting (B and C).

Carlsson et al.'s findings reveal that there is a continued reliance in the industry on formal exchanges that take place in structured, face-to-face project meetings. This is particularly so under traditional procurement regimes where the parties involved are likely to be fragmented and hence, separated by contractual and functional interfaces. It can be appreciated that as such barriers are likely to be broken down by more integrated project delivery systems, where the trust and understanding that develops is likely to place more of an emphasis on informal communication mechanisms. However, regardless of which procurement route is used, verbal exchanges in the form of meetings or telephone conversations form the cornerstone of interaction within the construction industry.

The importance of effective interpersonal communication for construction project performance

Effective interpersonal communication arguably lies at the heart of an effective organisational system. Indeed, without effective verbal exchanges, construction processes would be rendered impossible. Recent research undertaken in Israel (Shohet and Frydman, 2003) emphasises the importance of verbal communication to the success of projects. Their study focussed on communications in a construction management procurement of 30 residential projects. Informal communications were found to be crucial in ensuring the efficiency of the construction manager, with 48 per cent of their interaction being spent either in telephone conversations or in face-to-face meetings. The study showed that 'effective' projects were characterised by positive communications between the construction manager and the design team, in both the design and construction phases. It was also found that construction managers in effective projects devoted significantly more time (18 per cent compared to 4 per cent in 'ineffective' projects) to quality issues within their communications. This research underlines the importance of effective working based on appropriate methods of interpersonal communication.

An excellent case example of the importance of effective verbal interpersonal communication in the construction sector can be found in the experience of the Millennium Dome project, built in London during the late 1990's. The project team were housed in open plan offices that were intended to facilitate a 'binding process', helping the co-located team to work together more effectively. In this project it was intended that people interacted on a one-to-one basis, rather than rely on formal written media (*Contract Journal*, 1999a). By adopting this principle and breaking down the physical barriers created by locating members of the project team in remote areas, problems could be rapidly identified and addressed by relevant members of the team. Such an approach is well supported by research undertaken in other industries, which has shown that managers have a strong preference for verbal communication over other forms (Mintzberg *et al.*, 1976). Although the fact that face-to-face interaction is the most effective form of interpersonal communication in construction project environments, the need for co-location and/or physical meetings between the project participants presents a substantial problem for larger-scale projects, which are increasingly operated on a national or international basis. A continued reliance on direct interaction may require project participants to travel long distances and/or work away from home, which may have negative work/life balance implications. Thus, finding technological ways of facilitating one-to-one interaction has formed a focus of ICT developments in the industry in recent years (see Chapter 8).

An important corollary of the informal network which emerges in organisations and projects concerns the communication of humour amongst and between participants. This is important as it helps establish camaraderie in interpersonal relationships. Irony, sarcasm and wit communicated between colleagues are often considered to be 'banter' that is representative of the 'culture' of the construction industry. To strive for a wholly 'politically correct' working environment, would be to ignore the informal interaction that creates and sustains these temporary networks. Part of the rich construction dialect involves the use of foul language, sometimes manifested in the communication of humour. Indeed, in 1996 one trade publication reported that a site labourer won an industrial tribunal case for refusing to sign a no-swearing contract. Clearly the legal fraternity and those within the construction industry expect such expression onsite. Such behavioural expectations can have a negative effect on the way in which people communicate in construction, particularly in relation to the image that the industry portrays to external parties (see Table 6.2 in Chapter 6). For example, Riemer (1979) suggested that 'wolf whistling' at women passing by building sites occurs because operatives are expected to indulge in such behaviour. Thus, some verbal communication is not desirable in terms of improving industry performance.

Verbal interpersonal communication in construction

Although the act of face-to-face verbal communication may appear relatively natural and straightforward, the nature of today's construction industry, its fragmented structure and operational constraints, can all act as barriers to communication effectivity. Some of the constraints that these issues impose are discussed later.

Professional and process discontinuities

Construction projects comprise a rich network of different types of people (from unskilled labourers to highly skilled professionals) who are situated within different parts of the supply chain (from client advisors to facilities managers). Between them lie many interfaces defined by project position, contractual roles and obligations, societal expectations and stereotypes and individual nuances and traits. In many respects it is the multitude of people involved and the richness of communication that stems from this mix, as they interact within the project delivery process, that make it such an interesting industry to study from a communication perspective. However, such interfaces and differences in perspective can also lead to misunderstandings and/or barriers to communication which have to be overcome if projects are to be successful. These are explored later.

The professional language and semantics of construction communication

Given the vast numbers of people involved within the construction project delivery system, it can never be guaranteed that verbal exchanges are founded on common understanding. Different professionals and occupational groups within the industry have developed their own languages and communication processes which may not always be compatible with those with whom they work. A good example of this is provided by Moore and Dainty (1999, 2001), who carried out research into how the design and build procurement route could be undermined by issues arising from the rigid professional cultures of individual participants within project work groups. These have the potential to inhibit the achievement of a key-espoused benefit of this form of procurement; that it promotes the integration of the design and construction processes for improved project performance. Moore and Dainty's research showed that cultural non-interoperability acted as a significant potential barrier to effective change management, and the achievement of innovation within the design and construction processes. They argued that project responsibilities, which are currently delineated along professional identity lines, produce design and construction solutions that fail to fulfil the potential of integrated procurement. This research emphasises that those from different parts of the supply chain may speak 'different languages' insofar as their interpretation and understanding of roles and responsibility are rooted in their occupational role context. It is likely, for example, that an engineer will find it easier conversing with another engineer with whom he shares the same occupational parlance than with a landscape architect with whom he has no common professional language.

The increasing internationalisation of the construction industry's labour market and the implications for safe and effective working

An issue of increasing importance to the effectiveness of verbal interpersonal communication in the United Kingdom is the construction industry's reliance on the use of overseas labour. In recent times, the industry has remained relatively buoyant and it is now beset with skills shortages across its professional and craft workforce (CITB, 2004; Dainty et al., 2004). Accordingly, the industry has begun to draw upon a pool of migrant workers to offset its skills needs. With the European Union having recently granted accession to several new Eastern states, there is likely to be an even larger pool of potential skilled labour available to the UK market. Throughout 2002–2004, the construction press reported difficulties experienced on projects where non-English speaking personnel are employed. *Building* (2002a) for example, highlighted the challenges of having to translate Romanian, Farsi, Ukrainian, Punjabi and Albanian phrases into English. It is thought that around 19 per cent of construction operatives in London, for example, are foreign nationals (*Construction News*, 2003).

Clearly, a largely foreign workforce will present added problems for project managers in managing employees with limited proficiency in the indigenous language of their workplace (see Loosemore and Lee, 2002) and accommodating the cultural diversity that an influx of overseas workers will bring (Loosemore and Al Muslami, 1999). However, it also presents another problem for non-native English speakers themselves, in that no matter how good their language abilities, on construction sites verbal messages conveyed from sender to receiver are often augmented with additional verbal and non-verbal messages (e.g. tone and contradiction). Without this reinforcement to the verbal message, it is unlikely that the information will be recalled in the future. Whilst this has severe implications for employees' understanding of production information, it has potentially even more significance with regards to health and safety. The use of signage in the construction industry is generally pictorial in nature and used mainly to communicate safety information. As Lancaster *et al.* (2001) note, such pictorials are assumed to have universal information transmission potential. However, the comprehension of many signs may not be immediate to all employees within a project environment, particularly in projects with many different nationalities. Schellkens and Smith (2004) found that some construction site supervisors were unaware that the 1996 EC Safety Signs Directive required employers to explain unfamiliar signs to their employees.

As a means to prevent linguistic differences acting as a barrier to transmitting safety information, the Health and Safety Executive (HSE) and the Trade Union Congress (TUC) published a leaflet – 'Your health, your safety: A guide for workers'. The leaflet sets out workers' rights, what workers should expect from their employers and where to go for help. It is translated into some 21 languages which is a clear recognition that workers come from many diverse cultural backgrounds, and that all are entitled to access information and advice that meets their needs. The main construction trades union, UCATT, have also been active in this area with learning representatives helping migrant workers by running basic language courses.

The impact of personality and emotion in interpersonal communication

As was emphasised in Chapter 3, communication is a very personal thing in that alongside the interactive processes that characterise information transfer between people, a range of cognitive intrapersonal processes are also occurring in terms of individuals making sense of what has been transmitted to them. This emphasises the fact that different people think differently and hence, interpret dissimilar meaning into the same types of communication. Personality type therefore has an important role to play in determining the success of interpersonal communication and to how effective actors are in conveying their message to others.

Culp and Smith (2001) argue that individuals with different personalities approach an engineering project in different ways. They provide an interesting review of the Myers-Briggs Type indicator (MBTI) instrument, which is probably the best known instrument for measuring personality type. During the 1930s, Isabel Myers and Katharine Briggs developed a psychological instrument that was capable of measuring the observations made by noted psychiatrist C.J. Jung. Jung suggested that certain aspects of human behaviour are predictable and classifiable. This inspired Myers and Briggs to develop the tool based on four scales, with two opposite preferences defining the extremities or poles of each scale. Jung's theory suggests that everyone has a natural preference for one of the two poles on each of the four preference scales. A person may use both poles at different times, but not both at once and not with equal confidence. The four scales are:

- Extroversion or Introversion
- Sensing or Intuition
- Thinking or Feeling
- Judging or Perceiving.

Culp and Smith (2001) provide several examples which show how construction professionals can be characterised by these differences. From a communications perspective, the extrovert–introvert scale provides evidence of possible conflict. Extroverts prefer to communicate by talking whilst introverts prefer to communicate in writing. They cite one case study where a team of 18 comprised 2 extraverts and 16 introverts. They note that once the team members understood each other's style, team cohesion improved. Culp and Smith provide other examples from the remaining three scales and recommend its use as a means to understand and capitalise on the different behavioural styles within project teams. However, regardless of whether a sophisticated measurement tool is used, it is incumbent on all people attempting to communicate with others that they tailor their approach in such a way that it accords with their individual needs and personal frame of reference.

Birrell and McGarry (1991) suggest that a major step in improving the quality of communication in the building project team would be to establish the professional personality of each member of the core team. Once this was undertaken, guidance could be provided to participants regarding how best to communicate with each professional role. These researchers used the Kiersey Temperament Sorter tool (see Kiersey and Bates, 1984), a technique similar to the Myers-Briggs Type Indicator Test, which involves completing a personal choice questionnaire. Birrell and McGarry secured the cooperation of 55 construction professionals in the United States. The results showed that persons of similar personality temperaments might tend to cluster within the construction industry. Furthermore, the results suggested that these professionals were sensing–thinking–judging (STJ) types with tendencies towards extroversion. However, these types of individuals tend to be fact-orientated and so a team consisting solely of such individuals may have trouble resolving communication conflicts, particularly as each member is likely to be reluctant to compromise. Thus, personality types may also explain why interpersonal communication may fail, even where structural constraints grounded in the organisational project system are overcome through innovative procurement approaches.

Another important personality-related impediment to effective communication concerns the emotions of those involved in a construction process. If the environmental conditions that can exist within the construction industry are considered, particularly the need to complete projects within a given time frame, it is clear that the concepts of psychological stress and burnout are highly relevant to an individual's ability to process relevant information and respond to messages received (see Loosemore *et al.*, 2003). A high stress environment is likely to impinge upon an individual's abilities to make reasoned judgements as to the most appropriate communications channel and medium and may affect the clarity with which they convey the messages to their colleagues. The impact that emotions have on successful interpersonal communications has been investigated by Metts and Bowers (1994). They suggest that emotion is information that can be communicated verbally and/or non-verbally. Considering that it is virtually impossible to encounter a day that is free of emotion in construction project environments where people can be stressed, depressed, excited, happy, lonely etc., this is likely to impinge upon most aspects of interpersonal communication within the industry. Metts and Bowers argue that emotions can be felt but not expressed, expressed but not felt, intensified, attenuated, and sent both strategically and unintentionally. Thus, they suggest that emotions present a significant challenge for those who wish to study their role in interpersonal communication, as unless we understand their effect, the barriers to communication cannot really be fully understood. The role of these non-verbal messages in affecting interpersonal communications is explored in more detail later in this chapter.

Other impediments to the verbal communication process

A whole range of factors can adversely affect the effectiveness of communication between individuals that may be completely outside of the control of the individual sender of a message. This noise (see Chapter 3) may stem from the environment within which the communication is taking place (construction projects being characterised by activity and potential distraction), or from the nature of the individual being communicated with. This applies as much to non-verbal exchanges just as much as it does to face-to-face interaction. A study undertaken by Tam *et al.* (2003) within the Hong Kong construction industry revealed the substantial problems that exist in comprehending signs and symbols posted on construction sites. Their survey sought to uncover the relationship between the characteristics of construction personnel and their comprehension of mandatory action signs, prohibition signs and warning symbols. They found that several issues, including drinking habits, played a role in their respondents' ability to comprehend the meaning of safety symbols. Anecdotal evidence suggests that drugs and other stimulants can be found on some sites (Building, 2001b). An article in *Contract Journal* (2004b) reported that 12 per cent of operatives tested positive for drugs on one large construction project. An exhaustive review of the possible implications of this type of behaviour is clearly outside of the scope of this book, but these findings underline the fact that interpersonal communication can be affected by variables which would not necessarily be considered by managers of the construction process.

Non-verbal interpersonal communication in construction

As was discussed in Chapter 3, there is no shortage of research into non-verbal interpersonal communication. The study of non-verbal cues in communication is a truly interdisciplinary field with scholars from psychology, psychiatry, linguistics, sociology and management among those with an interest in this phenomenon (Knapp and Hall, 2002). Different studies attribute varying significance to non-verbal communication in shaping the communication process. A British Institute of Management report concluded that 10 per cent of a message is conveyed by intonation/expression and 40 per cent by body language (Pennington, 1986). Riggenbach (1986) suggests that more than 60 per cent of all communications is non-verbal. Regardless of which of these studies is to be believed, it is clear that non-verbal cues have a major role in face-to-face interaction. Their role in interpersonal construction communication is explored in more detail later.

Body language

As was discussed in Chapter 3, body language generally refers to unconscious non-verbal cues that people have little or no control over. As such, they can contradict or neutralise what is being said. It is important that managers gain an appreciation of non-verbal cues so that that they can put their messages across effectively and correctly interpret feedback from those with whom they have direct contact. Considering that construction projects are based around a series of formal and informal interactions, which together determine the success of the production effort, body language may be a major determinant of the success of construction projects.

Some processes within construction can be seen to be more dependent on non-verbal communication than others. For example, the process of negotiation regularly occurs between parties from different parts of the supply chain. The physical gestures made during the process of interpersonal communication reflect the inner attitudes of the parties involved and their ability to read minute gestures will assist in negotiating effectively (Riggenbach, 1986). These may not always accord with what is actually said in such negotiation meetings. Riggenbach (ibid.) suggests that several factors listed below need to be acknowledged when analysing body language:

- it is learned, and hence varies from country to country, culture to culture;
- business body language and social body language can vary;
- it is a subconscious reaction and tells more than the spoken word;

- it shows the inner feelings and attitudes of a person (actions speak louder than words); and
- understanding the situation is the key to analysing any gesture, and one gesture can have several interpretations depending on the circumstances.

These features of body language render it a problematic area of communication to understand, particularly given the surrounding constraints provided by the construction project environment. From a construction industry perspective, no specific training is given on how to read body language within undergraduate- and postgraduate-built environment courses. It could therefore be argued, that few within the industry are likely to consider the complete 'message' in non-verbal communications other than the most blatant emphasis in voice and tone. Although experience of interaction is likely to inform an individual of some of the feeling surrounding a particular interaction, a fuller understanding of such issues is likely to be beneficial in understanding other parties' needs and resolving conflicts within construction project environments.

Within the context of international construction, Howes and Tah (2003) suggest that non-verbal cues can be used advantageously, or can give serious offence. This reflects the fact the same symbolic gesture can have different meanings in different countries. They cite Stallworthy and Kharbanda (1985) who note that the vertical movement of the head (nod) means 'yes' in most countries, but in Greece and Turkey it can mean 'no'. This type of anthropological knowledge has been employed by the HSBC bank in their worldwide marketing campaign to stress the importance of regional 'local' knowledge when doing business in other cultures (HSBC, 2004). Considering the increasing use of foreign labour on UK construction projects and the shift towards a more globally defined construction market (as was alluded to in Chapter 2), it is likely that these gestures will play a more significant role in the efficacy of interpersonal communication in the industry in the future.

Other non-verbal cues

As was discussed in Chapter 3, non-verbal cues are not restricted to the unconscious gestures that people make, but can also be manifested in the way in which people dress and behave within the workplace. For example, professionals and craft operatives employed in the construction industry often communicate their occupation role/identity by way of costume and jargon, an implicit indicator of occupational socialisation (Riemer, 1979). Visual stereotypes of professional occupations in construction include the architect with a cravat, the quantity surveyor with desk clerks armbands or a project manager with gold cufflinks! Although these stereotypes are likely to be discounted by many working in the industry, they can limit communication between certain parties according to their perceived role (see Loosemore and Tan, 2000). In the United Kingdom, trades such as plastering and painting and decorating are associated with white 'bib & brace' overalls, whilst bricklayers, electricians, carpenters and plumbers are not always seen wearing recognisable attire. The purpose of clothing does of course differ from that of the professional in that work wear provides protection from the elements and dirt and can be used to carry tools and supplies. Overalls also provide potential advertising space for displaying a corporate logo (see Chapter 7), or perhaps more informally an ability to perform a manual job. Unfortunately, the communication of occupational role through dress often backfires. For example, the term 'builders-bum' is now somewhat institutionalised in the United Kingdom and used by the media to characterise overweight builders with low slung jeans. This differs markedly from early twenteith century images of Edwardian tradesmen with dress jackets, waistcoats, ties and bowler hats.

The interplay of verbal and non-verbal forms of interpersonal communication

As was briefly discussed earlier in this chapter, interpersonal communication does not occur in a vacuum, but is reliant on a combination of verbal and non-verbal cues which occur concurrently. It is the combination of verbal and non-verbal cues that leads to understanding in successful interpersonal communication. Indeed, it is a generally applicable rule in interpersonal communication that using a variety of different media reinforces the message, facilitates decoding and hence, enhances understanding. Richmond *et al.* (1991) describe the function of non-verbal messages in relation to verbal messages in communication. They suggest that six functions are relevant in describing the interrelationship between verbal and non-verbal communication as listed in Table 4.2. This table also provides examples of how such non-verbal communication can be manifested in construction.

Richmond *et al.*'s functions appear to have a close relationship with the seminal research undertaken by Bales (1950) who developed an Interaction Process Analysis (IPA) tool. The IPA tool allows researchers to observe face-to-face communications within group settings and classify their interaction into one of twelve sub-categories. The purpose is to tease out two distinct categories of interaction, either socio-emotional or task-based interaction. The IPA model has been used widely in research by construction researchers in order to understand the behaviour manifested in such interactions (see Gameson, 1992). Such an approach allows researchers to begin to understand how project decisions are made and where problems and discontinuities may lie within the communication process.

In addition to the combination of these deliberate and non-deliberate behavioural gestures, more formal supporting cues also have a role to play

Table 4.2 Function of non-verbal communication in relation to verbal messages

Function	Description	Construction example
Complementing	Adds to and reinforces or clarifies the meaning of a verbal message, that is, use of a pleasant voice to convey meaning	Project manager quells potentially adversarial situation at project meeting by *calmly* explaining difficult situation to relevant parties
Contradicting	Instead of complementing the verbal message, some non-verbal messages contradict the verbal, that is, the use of sarcasm to convey the opposite message	Architect indicates that he is happy to stay on for extra time at meeting but continually checks his watch thereafter
Repeating	Non-verbal message repeats the verbal, and vice versa, that is, the use of a 2-fingered v signal to denote victory	Banksman gives the 'thumbs up' to crane driver to reinforce his verbal message of 'OK'
Regulating	Control of verbal interaction with non-verbal messages, that is, looking at or away from the other person, raising a finger while pausing to indicate you are not finished	Site safety officer reprimanding an employee who starts to turn away by gesturing to him to listen to the instruction being given
Substituting	Used when non-verbal messages are sent instead of verbal messages, that is, waiving or beckoning another person	Hand instructions given by precast concrete operative to crane operator
Accenting	Non-verbal message used to accent or emphasise a verbal message, that is, saying something louder than normal tends to emphasise the verbal message, touching someone while talking tends to emphasise what is said	Site manager pats commercial manager on the back congratulating him for a skilful piece of negotiation in relation to the settlement of a sub-contract package

Source: Cf. Richmond *et al.* (1991).

in reinforcing verbal communication within the construction industry. Table 4.3 shows the results from a study undertaken by Wogalter and Sojourner (1997). It shows how a combination of verbal instruction and visual stimulation more effectively reinforces understanding of a message in both the short and longer term. These results are highly significant in construction, where interpersonal interaction may involve transmitting large amounts of verbal information very rapidly between project participants. Consider, for example, the importance of compulsory site safety inductions that now take place on most large sites. Here, ensuring that those being inducted fully understand the health and safety information is crucial, which underlines the need to reinforce verbal information within visual

Table 4.3 Percentage recall of information transmitted in different forms

Method of conveying information	3hrs later (%)	3 days later (%)	3 weeks later (%)	3 months later (%)
Telling and showing together	85	65	28	16
Showing (just pictures)	72	20	12	7
Telling (just voices)	70	10	7	3

Source: Wogalter and Sojourner (1997) cited in Lancaster (2001).

messages intended to influence behaviour. During the construction of the Hoover Dam project in the United States, the contractor initiated a well-publicised safety programme that included billboards around the site with a bold black message which read '*Death is so Permanent*' (Stevens, 1988).

Reinforcing a message in different ways is particularly important in change situations where people will need to hear about the change several times for them to absorb it and understand the implications in terms of their own work (Weiss, 2000). Weiss suggests that effective communication during change implementation requires the following approaches:

- Multiple communication media should be chosen to influence people to accept the change – that is, a combination of written, spoken, electronic, experimental and educational media.
- A specific time and medium should be identified for regular communication – when people know that communication is coming they can adjust to change more effectively.
- Every action should be linked to the desired future scenario following the change – this helps people to understand why the change is being implemented and reinforces the anticipated results.
- Successes should be identified and celebrated as they occur – highlighting success helps them to motivate people to readily accept the change.
- Expectations of the speed and quality with which the change will be implemented should be managed – change leaders must be open and honest in their communication in terms of what they can realistically achieve.

This strategy emphasises the importance of leaders using mixed methods of communication and reinforcement tactics to ensure that their message is understood and appropriately acted upon. Project managers will often send memos or emails to underscore the importance of a particular issue and ensure that his/her staff act upon their instructions in response to it. The ways in which project managers can do this within teams is explored in more depth within Chapter 5.

Summary

Construction is a people-intensive, highly social industry where people interact regularly in support of the production effort. The face-to-face channels used to exchange information throughout the project lifecycle is known as interpersonal communication, the efficacy of which to some extent defines the success or otherwise of construction projects. In this chapter, the nature of interpersonal communication has been explained, together with how it tends to be manifested within the construction project environment. This has revealed the reliance within the sector on face-to-face verbal communication as the primary mechanism for individual interaction between project participants. However, there remain many barriers to effective communication grounded in the fragmentation of the industry's procurement and production processes, the increasingly diverse nature of the sector's labour market and the nuances of human behaviour, which together render common understanding of meaning very problematic within, the construction project environment. Of particular importance in understanding interpersonal communication within the industry is the role of body language and other non-verbal cues which add meaning and depth to verbal exchanges. Such behaviourally determined dimensions of communication are complex, but project managers must attempt to understand them if they are to ensure that their messages are being understood by those around them. The starting point for understanding the process of interpersonal communication is, perhaps, an acknowledgement of other peoples' intrapersonal interpretation and perspective. Thus, understanding the personal beliefs and culture of the intended recipient will help to determine the appropriate combination of verbal communication and non-verbal cues that can be used to facilitate understanding. The skilled construction project manager will do this both consciously and subconsciously throughout their day-to-day interactions with those with whom they interface. Learning to strategically combine verbal and non-verbal cues in order to manage people more effectively is a crucial leadership quality, particularly in a people-centred industry such as construction.

Critical discussion question

1 Consider a meeting between a contractor's project quantity surveyor and an owner/manager of a subcontractor with the aim of settling the final account package. The work package has run over time and budget due to coordination problems with other work packages that were

largely outside of control of the main contractor. This presents a potentially confrontational situation between the parties involved.

- Discuss the ways in which emotions and feelings may be expressed by the parties during the meeting.
- Advise the quantity surveyor on the non-verbal cues that he may notice on behalf of the subcontractor should he feel aggrieved at the negotiated settlement that he is being offered in respect of the works.
- Outline a personal communication strategy for the quantity surveyor that will help him/her to maintain a positive dialogue with the subcontractor and avoid a potentially damaging dispute situation.

Case study: inter-professional communication

Introduction

The concept of interpersonal communication was introduced earlier in this chapter as being communication between two or more individuals. This definition can be viewed as valid in a generic context in that it does not specify any of the characteristics of the nature of the communication between the individuals involved. Within the construction industry there are, however a considerable number of distinct professional groups who form what could be referred to as 'tribes'. The members of each professional group share in common some modes of behaviour, along with values and beliefs, but of more importance to this chapter they also share a common 'language'. The term 'language' in this context does not refer to the fact that everyone in a group speaks English (or Welsh, or French, etc.), but to their specialist technical language. This will include unique terms and phrases that are used to communicate the actions which relate to the particular functional specialism of the group in question. Within construction, these might include bricklaying, joinery, plumbing, project management, quantity surveying, etc. This case study begins by presenting a critical discourse around professional and functional demarcations within the industry, before exploring how such fragmentation is manifested within the context of a large, technically complex project.

Professional and functional communication boundaries

There are a considerable number of specialisms within the modern construction industry and all of them have to be integrated within the boundaries of whatever project is to be implemented. This fragmented picture is rooted in the disparate skills, expertise and knowledge of the individual members of an individual specialism (or tribe). In the context of planning the execution of a construction project, such a consideration is entirely appropriate in that there is a need to know when one work package is completed (and its attendant specialist workforce is no longer required) and the next has to commence (therefore requiring a different specialist workforce to engage in the project). However, considering specialist groups in this manner does not explicitly address the issue of communication, whether that may be at the inter-specialism or the intra-specialism level. In the context of work planning this is not a problem. However, when managing the real-time implementation of the work as planned it can become a considerable constraint on performance. Arguably this problem has the greatest potential to manifest itself as a disintegrative force when considering the process of communication at the level of professionals (rather than craft specialisms).

There are two reasons for suggesting that communication has the greatest potential to be a disintegrative force within the context of inter- and intra-professional relationships. First, the activities of professional groups do not directly result in the production of any component or sub-assembly of the building or structure that is the focus of the project. The production process is indirect in that there will either be a manual construction process and/or a manual/automated manufacturing process between the output of any professional and the integration of a component or subassembly into the fabric of a building. Such a situation is inevitable given the level of fragmentation that has come about during the evolution of the construction industry. Intermediate processes may be seen as being beneficial, in so far as they are a buffer between the professional output (i.e. the design) and the onsite production activity. Should there be a problem with the professional output, there is potentially an opportunity to remedy that problem prior to commencing related work onsite. In the planning context this is arguably so, but in the communication context such a buffer becomes an additional interface as a third party becomes involved in the communication process. The third party may add a further problem in that they may not be 'fluent' in the language of either the professionals whose output they are reliant upon or of the onsite operatives who will be responsible for integrating their product into the fabric of the structure. The second reason for the disintegrative potential relates to the complex language used by professionals within the industry. Examination of almost any standard form of contract will serve as a ready illustration of the point regarding complex language. The tradition of increasing functional specialism has had as significant an impact on the level of fragmentation within the professional sector of the industry as on the manual sector, and this has arguably produced more communication problems than for the construction craft occupations. This impact of this professional (or tribal) language is explored in relation to the case study later.

The case vignette project

As was stated earlier in this chapter, there is evidence that effective interpersonal communication is vital for the success of the industry and that verbal interaction forms the cornerstone of interaction within the industry. On this basis, the analysis of interpersonal communication between professional groups within a sample construction project represents a valid case vignette for this chapter. The project to be analysed is a multi-million pound distribution centre procured under an integrated design and build (D&B) process. The stated objective of the client in utilising this approach was to maximise the integrative nature of the system, removing interfaces within the design/construction delivery chain and optimising the probability of delivering the project on time and within budget. The appointed contractor (a large UK-based firm with in-house design and construction resources) was well aware of the fragmented nature of the communication and interaction between the various professional groups employed by them. They also saw D&B as a possible means of building a fully integrated project team which could overcome such fragmentation, and had worked actively to expand this area of their business. They directly employed all of the key players involved in the design and construction of the project, including design managers, architects, structural and building services engineers, quantity surveyors and construction management staff. The work packages were procured externally through a group of specialist subcontractors, most of which had worked for the company previously and had achieved preferred subcontractor status. Thus, in principle, the delivery of the project should have been a relatively seamless operation for which to deliver the project.

Fragmentation within an integrated system?

The contractor's approach to building its 'integrative' team was an uncomplicated one: they merely placed all the team specialists required for the project into one room and left them to get on with it. There seemed to be an assumption that the individuals involved would be sufficiently 'professional' to recognise that the success of the project depended on them all working in an integrative manner and therefore start integrating. Unfortunately, people are a little more complex and infinitely more intransigent than the contractor appeared to have believed to be the case. Many of the project participants had relatively little experience of working in a D&B environment prior to this project. Indeed, many had no experience of integrated environments and were therefore unsure of how to proceed. This uncertainty was evidenced in the initial structuring that the team adopted, where each professional group adopted a cluster of adjacent desks. The result of this was that the project team immediately structured itself within its professional groups. In effect, they had each withdrawn to the safety of their professional silo.

Silo structures are an outcome of traditional, transactional management approaches. It was established shortly after the Industrial Revolution that with increasing specialism comes greater efficiency in production. This resulted in increasing levels of differentiation (based on either time, technology or territory) within the production process for all mass-produced articles. This differentiation was managed on the basis of silos in which specialists worked in a manner largely devoid of communication with other silos involved in contributing to the production of the same product. In an environment characterised by slow rates of change, such a management approach presents more benefits than disbenefits. However, today's construction industry has long since ceased to operate in such an environment. Nonetheless, it can be argued that the industry is still characterised by silos of professional activity. Observers from outside the United Kingdom will often comment on how many professional bodies there are providing governance for the industry. Each of these organisations will seek to protect the interests of their members by laying claim to a particular knowledge 'territory' that differentiates them from others. Arguably, this professional protectionism acts as a barrier to change, but it also acts as a barrier to communication.

The implications of disintegration in the project team

In the case vignette project the professional groups maintained their silos both within the work environment and in their leisure time and there was little evidence of work or social interaction between individuals in different groups during the early stages of the project. Given that verbal interaction is the cornerstone of communication within the construction industry, such a withdrawal from potential opportunities for verbal interaction represented a significant concern for the development of an integrated project team. However, team life cycle models typically refer to early problems of this kind, with terms such as 'entrenchment' being used to describe the behaviour of 'protecting' values and beliefs when they may potentially be challenged by new team members. Whilst entrenchment should therefore be expected to be inevitable (in some researcher's opinions it is in fact desirable) it should, if the team life cycle unfolds naturally, only represent one stage of several. If, however, the life cycle is not completed (the process becomes stuck at one stage) then what is formed cannot be considered a team. It remains, at best, a group with an attendant level of performance less than that of a team.

The case vignette project 'team' did in fact become stuck, with individuals actually seeking to avoid face-to-face verbal interaction with certain other individuals. One extreme example was a member of the architect group who refused to speak to one member of the construction management group. The only way that verbal information could pass between these two individuals was for it to be conveyed by a third individual who was from the Quantity Surveyor (QS) group. By largely eliminating verbal interaction the group members' main communication media became the non-verbal forms such as email, faxes and even letters. This is particularly ironic in that the main positive result of the contractor's decision to place all the team members in one location was that it had achieved possibly the most desirable state for large teams on complex projects: physical co-location. Teams who are deprived of this state invariably complain about the problems of having to communicate without the benefits of actually being able to go and talk face-to-face with other team members. The case vignette project team members were largely seeking to dislocate themselves within an ideal (co-located) environment and preferring to communicate through relatively barren media such as emails (the problems of communication using IT are discussed in Chapter 8). Consequently, their initial level of performance was no greater (and possibly less) than would have been achieved by a traditional, non-integrated team. One member of the construction management group remained convinced at the project's completion that if the communication problems had been solved early in the project, the contractor would have been able to deliver the completed structure ahead of schedule and considerably below budget. This would have resulted in a more favourable profit situation as the contract allowed for a sharing of savings between the contractor and the client.

Communication resolution

At a relatively late stage of the project it became evident that the project groups had become, to a greater extent, unstuck. The level of verbal interaction (both in the work and social environments) increased considerably as individuals moved out of their silo mentality. Consequently, both the rate at which problems were solved and the quality of the solutions applied, increased considerably. Without this late improvement in performance it is almost certain that the project would have been completed behind schedule and over budget. The contractor appeared to regard this late surge as evidence that their integrated project team approach had worked successfully. It is true that there was evidence of integrated working between team members in the latter stages of the project, but little of this was achieved through the actions of the contractor. Nor was it achieved by applying more effective ways of using information communication technologies. The most significant factor was the changing of the project manager, with the new project manager recognising that the supposedly integrated team was actually largely disintegrated and then taking action to try and redress the situation. The actions taken were targeted at encouraging groups to take a less confrontational approach to each other through not focusing on their 'professional' relationship within the contract and through that seeking to impose a traditional hierarchy. In this manner, the work environment improved quite rapidly, thereby allowing the late surge of increased performance.

Conclusions and lessons learned

This case vignette reveals the importance of language and culture to the effective integration of construction project team members. Communication between individuals lies at the heart of interaction between members of the construction team. However, no matter how structurally integrated a group may appear to be, interpersonal interaction cannot take place unless there are concurrent efforts to break down the barriers which impinge upon effective communication among them. Overcoming professional and functional boundaries demands that those involved work to develop a common language of understanding, taking time to address any difficulties in understanding and appreciating where misunderstandings and barriers to interpretation may occur. Clearly, within a construction context, the role of project manager is fundamental to ensuring that such barriers are overcome. In many respects they can be seen as the facilitator of the communication process between individuals within their team. They must, therefore, be acutely aware of any tribalism or silo mentalities building up within their team.

Chapter 5

Group and team communication

Construction project activity is a collectivist endeavour. This means that it will inevitably involve groups of people with different skills, knowledge and abilities working together, who each will ideally make a distinct contribution to the overall production activity. For example, consider a project team managing the construction of a large building for a contracting organisation. This will involve a range of construction managers with responsibility for the various packages of work, quantity surveyors overseeing the commercial aspects of the endeavour, site engineers responsible for setting out the works, first line supervisors dealing directly with the workforce and subcontractors and a project manager overseeing and coordinating the work of the team and maintaining the relationship with the designers and client organisations. Each person will have distinct responsibilities, but these will support and complement those of their colleagues. Groups of people working together in this way are often referred to as being members of a 'team', which infers that they work together in a way that synergistically utilises their skills and knowledge. However, if they fail to communicate effectively, then they will be unable to exploit their collective talents and could instead operate less effectively as a disparate 'work group'. In this chapter, the way in which people work together in groups and teams within the industry is explored. Initially, internal team communication dynamics is considered, before exploring inter-group communication in terms of how teams communicate with their external environment.

Group development and team roles

The formation and development of groups and teams of people have provided one of the most significant themes in management literature throughout the latter half of the twentieth century. Understanding the ways in which teams form and develop is fundamental to understanding how they communicate within the construction sector.

Group formation and development

The interdisciplinary nature of construction project teams is such that they will always involve people from different organisations and with different backgrounds, skills and knowledge coming together for short periods of time to work together. The process of group formation and reaching a stage where people communicate effectively (or 'speak each others' language') can be problematic in an industry characterised by fragmentation and temporary involvement on behalf of its main participants. Nevertheless, as is explored later in this chapter, effective group work depends upon those involved working effectively within their particular team roles. These will be variously defined according to both their formal roles and their individual skills and contribution. Initially, therefore, in order to understand the communication process within groups it is first necessary to understand the ways in which groups form and develop in construction and the impacts that this has on communication processes within the project environment.

As a group of individuals work together on the same or on connected tasks, they begin to develop as a work group and ultimately become a 'team'. This infers that members will have defined their individual roles and contribution to the activities of the group in a way that best utilises their skills, knowledge and abilities. Although there are difficulties in developing teamworking within the industry, it is still evident that in many projects a positive working environment evolves that leads to successful outcomes for the parties involved. In such projects, individuals have managed to engage in socio-emotional interaction in such a way that they have been able to maintain relationships and coordinate their collective activities (Emmitt and Gorse, 2003: 60). Such understanding can stem from a number of events occurring in the lifetime of a team. For example, as the team resolve problems together, achieve project targets in terms of interim time and cost milestones and socially develop in terms of their informal interaction both inside and outside of work, they will inevitably become more confident with working with each other (i.e. trust will develop) and will learn to compensate for each other's weaknesses and exploit each others' strengths. A useful model for understanding this process is provided by Tuckman's (1965) model of group development. This suggests that there are four stages of group development as follows:

1 *Forming* – This is the early stage of a group's development where members get to know each other and try to gain acceptance into the group. Group behaviour is likely to be inhibited as members are likely to be careful not to cause conflict or offence to group members.
2 *Storming* – As individuals begin to gain confidence within the group setting they begin to feel more secure and state their opinions and perspectives more assertively. This inevitably results in conflict but will also raise awareness amongst group members of where potential problems within the group are likely to occur.
3 *Norming* – When the group begin to develop a sense of how they fit in and what their individual roles and responsibilities are going to be, group norms emerge which effectively structure interaction between

group members. It is at this stage when communication protocols will emerge and group cohesion can take place.

4   *Performing* – The group's norms provide a framework for effective working which help the group to work more synergistically.

Thus, when a team does develop successfully and the members begin to work to each others' strengths and compensate for each others' weaknesses, a degree of team 'synergy' can be said to have occurred. This infers that members have become comfortable with their own role and position in the team and can manage their individual involvement and contribution effectively. In an interesting study undertaken by the Tavistock Institute, Nicolini (2001, 2002) explored the concept of 'project chemistry' within the construction industry. Nicolini used focus groups to explore what was required to create 'good project chemistry' leading to successful project outcomes. He found that the respondents tended to associate it with good and open communication between members of the project. Understanding how to engender team synergy and develop positive project chemistry demands that members understand their roles and contribution in relation to those of their colleagues and communicate effectively between them.

Although the development of team synergy and a positive project chemistry is theoretically straightforward, the dynamic nature and characteristics of the industry and its projects mean that group development may be more problematic than in more static environments, as people will join teams for defined periods of time before moving on to other workgroups. This temporal dimension defines interaction and communication within the construction project team environment and renders it as amongst the most complex of all industries. It can have a fundamental effect on the way in which group members communicate with each other and hence, how the group develops through the lifetime of a project.

Moreland and Levine (2002) suggest that groups contain established (full) and new (marginal) members who are in different membership phases as they proceed through the socialisation process. This raises an issue of how much the full members can trust the marginal members, who, while they are classed as belonging to the group, are not yet fully accepted by it. The number of marginal members in any group will vary over time and the transition of members between two or more groups will also be variable. There is also the need to consider multiple group/team memberships as individuals may well be a member of more than one group. Such a situation may well affect the process of a member moving from marginal to established status within one group. There is also the added complication that an individual may be an established member in one group, but a marginal member in another. Such factors will impact differentially on the communication process, with an individual possibly communicating in different styles across the various project environments. A marginal member may well use a deferential style, whereas an established member may be allowed to use a more authoritative style. The overall group can then be argued to decrease in homogeneity and increase in diversity resulting in a limited common basis for the encoding and decoding of any given message.

Other factors can also inhibit group development within the construction project environment. The fragmented and temporary nature of construction teams defines an involvement climate in which people may not have a vested interest in the *overall* success of the project as a whole. For example, a subcontractor seeking to maximise profit on their package of work may not work effectively with other subcontract firms whose packages interface with theirs if it is not in their personal interests to do so. Thus, it is important to recognise that some members of a work group may not wish to work effectively as part of the project team, but would rather retain a distinct member of the project work group.

A final issue to consider in group development is that all teams will eventually function at a level below what they did in the performing stage of their life cycle. Tuckman added in another stage to his model known as the 'mourning stage'. Here, the team's levels of activities subside, members leave and remaining members may feel a sense of loss. Within construction, such feelings are likely to be frequently experienced as members' involvement will move on to other endeavours. The communication channels and protocols that have been developed within a successful project team will not necessarily work in another context. This places considerable demands on team members (and particularly the project manager) to redevelop these from project to project.

Team roles

Within all group and team situations, members will take on particular tasks and responsibilities. Some of these will be formally determined by position and occupational role, whereas others will evolve as each individual applies his/her particular skills and personality traits in a way that is beneficial to the achievement of team goals. They will undertake these tasks in ways which are, to some extent at least, determined by their behavioural contribution to team tasks. Probably, the most well-known classification of team-role types is provided by Belbin (2004). The Belbin Team-Role Self-Perception Inventory is a toolkit that can be used to assess the suitability of personalities in relation to team roles. During a period of over nine years, Meredith Belbin and his team of researchers based at Henley Management College, England, studied the behaviour of managers from all over the world. Managers taking part in the study completed a range of psychometric tests and were put into teams of varying composition while they were engaged in a complex management exercise. Their different core personality traits, intellectual styles and behaviours were assessed during the exercise. As time progressed, different clusters of behaviour were identified as underlying the success of the teams. These successful clusters of behaviour were then given names, which can be broadly classified under three broad headings:

- Action-oriented roles: shaper, implementer, and completer finisher.
- People-oriented roles: coordinator, team worker and resource investigator.
- Cerebral roles: plant, monitor evaluator and specialist.

The team role types describe a pattern of behaviour that characterises a person's behaviour in relationship to others' in facilitating the progress of a team. The value of Belbin's team-role theory therefore lies in enabling an individual or team to benefit from self-knowledge and adjust their contribution according to the demands being made by the external situation. Some of the team role types are beneficial or problematic to ensuring effective communications within a team. For example, a weakness in the 'Plant' role (those who are very creative and imaginative and like to solve difficult problems), is that they may be too preoccupied to bother communicating effectively. The 'Resource Investigator' is said to be an extrovert, enthusiastic and very communicative whilst the 'Teamworker' is cooperative and listens attentively. It can be appreciated, therefore, that a team dominated by 'Plants' could stifle communication between members whereas a group of 'Resource investigators' are likely to suffer from too many people speaking at the same time unless offset by a number of 'Teamworkers'.

Understanding personality types and team contributions may be key in understanding where team communication difficulties arise. This may help in both the selection of appropriate team members (Gardiner, 1993) or to removing members who do not fulfil a role or who adversely affect intragroup communications (Moxley, 1991). In a construction context, the Belbin approach has been applied to the development of project teams. However, such an approach is arguably more useful when exploring the synergies of the overall project delivery team in terms of the various stakeholders involved with a project endeavour. For example, if a contractor's team was appointed based on assessment of how their team was likely to work with the client's team, this could potentially reduce conflict and enhance workplace relations within construction projects. However, as Cornick and Mather (1999) recognise, it would be highly unlikely for a client's project manager to convince a client that a particular firm or individual(s) (who comply with all pre-qualification criteria) should not be appointed because of a risk of personality clash with another team member, firm or individual! Indeed, Blockley and Godfrey (2000) acknowledge that it is unlikely that a construction team could be designed to incorporate all of Belbin's profiles.

Factors shaping the success of communication within construction project teams

As has been alluded to above, construction projects are by no means simple and straightforward environments within which to ensure effective communication, particularly considering the temporary involvement that characterises their formation and development. The discourse within this chapter has thus far suggested that teams will naturally develop, that synergies will emerge and that they will function effectively. In reality, however, teams often fail, certain members do not contribute effectively and communication can break down, sometimes irrevocably. The reasons for this are grounded in the nature of group interaction, individual interpretation and the structural and cultural boundaries defined by the particular situation in hand. In order to understand the nature of intra-group communication further, it is necessary to explore the nature of interactions within such groups and roles and how positions can influence their interaction with other members.

The influence of formality in group interactions

In broad terms, groups of individuals can be said to interact on a formal or informal basis, depending upon whether they have been deliberately created (such as in the formation of a construction project team) or have developed themselves in response to a prevailing situation or set of circumstances (such as members of different organisations coming together to resolve a problem that has occurred on site). The formality of teams, and the nature of the way in which they have developed and evolved, has important implications for the way in which they communicate. According to Furnham (1997) *informally* defined groups are characterised by a spontaneously developed structure which has dynamic qualities. In contrast, *formal* groups tend to have a planned structure which tends to be more stable in nature. From a communication perspective, whereas formal groups are likely to use formalised communication channels, informal groups are likely to use similarly informal channels, such as the 'grapevine' to communicate between themselves. The entire basis upon which interaction between members is founded fundamentally differs therefore, with formal teams relying upon interaction prescribed by functional duties or position, whereas informal group interaction is based upon personal characteristics and status.

Within construction, the formality of interaction between group members can have a marked influence on their intra-group performance. A good example of this is provided by Loosemore's (1996) study of four traditionally procured building projects. This revealed how groups reacted to crises occurring on construction projects. He found that contract documents became more important as a formal guide to responsibility patterns, but differences in interpretation and understanding of the contracts was also common. With regards to communication, Loosemore (ibid.) found that there was a tendency for project participants to exhibit extremes of formal and informal behaviour. These situations apparently led to projects moving towards a downward spiral of poor communication, tension, anxiety and stress. Loosemore concluded that contractual procedures seemed to impose inappropriate formality in a crisis situation that actually stymies their ability to take action. The argument is that what is in fact needed during such events is 'informality and flexibility' and Loosemore proposes the inclusion of an 'emergency clause' within contracts to be enacted during a crisis. This clause would permit the parties concerned to resolve their differences in an informal manner and thus permit a temporary cessation to the formal, procedural, contract requirements.

The influence of roles and positions within intra-group construction communication

The roles that individuals have within a construction project team, whether they are formally or informally defined, will affect their abilities to communicate effectively. This is because barriers or boundaries will impede open communications between team members. These can be formally defined according to functional or hierarchical position, or more informally in terms of personality types or social status within the group. The impact that such boundaries have on communication within construction teams has been the subject of much research in recent years, and these studies have provided insights into how these roles affect the efficacy of team working within the sector.

Foley and Macmillan (2005) explored the dominance that certain professions have over interpersonal communication during project meetings. These researchers monitored four types of meetings (team progress, technical, interim technical and strategy/problem solving) during the construction phase of a new exhibition centre. A simple coding system was adopted for recording the nature and extent of the exchanges among team members. The objective was to measure the volume of communication between eight participants (contractor, architect, structural engineer, quantity surveyor, project manager, client, funding body and subcontractor). They found that three main players (the contractor, architect and project manager) formed a triangulation of interaction that contributed to 78 per cent of the verbal communication observed. Table 5.1 shows the communication input for the four meetings which formed the focus of the study. This shows the domination by the three main players was due to the discussions revolving around construction issues. However, it was found that the three main parties formed a core group acting as agents for some of the others – the contractor for example, communicated on behalf of subcontractors, and the user/client tended to communicate

Table 5.1 Communication input

Team member discipline	Progress meetings (%)	Technical meetings (%)	Interim technical meetings (%)	Strategy meetings (%)	All meetings (%)
Contractor	46	37	29	27	40
Project manager	15	14	15	21	17
Architect	16	33	37	16	21
Structural engineer	9	7	6	18	10
Quantity surveyor	8	5	6	2	6
User/client	3	3	0	9	3
Funding body	1	1	0	0	1
Sub-contractor	1	0	7	7	2
Other	1	0	0	0	0
Total time observed in meetings	14 hr 25 min	2 hr 55 min	2 hr 40 min	4 hr	24 hr

Source: Foley and Macmillan (2005).

through other members of the project team. Thus, many of the interactions between the parties were found to be indirect and outside of the formal project structure. This pattern was observed in all meeting types.

In another study, Murray (2003b) interviewed over sixty construction professionals engaged on eleven projects to uncover their perceptions about communication and decision-making behaviour. The six professions (Architect, Main Contractor, Project Quantity Surveyor, Mechanical and Electrical Engineer, Project Manager and Structural Engineer) on each project were requested to complete a questionnaire related to a series of project incidents that had occurred. The questionnaire required them to indicate who they communicated with and who they regarded as the main decision-maker in resolving the incident. The results revealed a distinct profile of communication and decision-making behaviour within the eleven case study projects. The research showed that some 57 per cent of the actors were not central to the communication networks and 49 per cent were not involved to any significant extent with the decision-making process. The results say much about the culture of the industry and why communications can be so problematic in team environments.

Both Foley and Macmillan and Murray's findings have resonance for the efficacy of communication in project teams comprising members from different organisations. The fact that certain individuals were acting in agency for others, or were apparently communicating through other parties, increases the likelihood of 'noise' affecting the transfer of information travelling from one person to another. It also results in more communication interfaces through which communication must flow.

Intra-team boundaries as inhibitors of team communication performance

The industry's basis of communication within an industrial context has largely evolved in response to the productivity strategy of increasing specialisation. A significant implication of specialisation, however, is that every new function defined results in a further boundary to communication which emerges from the new roles created. Even within (supposedly) integrated project teams, such interfaces can lead to discontinuities and barriers to effective working which can have a severe effect upon the team's abilities to react to situations. A good example of this can be found in Murray's (2004) account of the Sydney Opera House project. Murray's book provides a noteworthy account of the Opera House project which has some similarities with the New Scottish Parliament project (see case study in Chapter 2). Both were public sector projects with a 'fast-track' procurement approach, employed signature architects, involved complex engineering and suffered from communication problems which eventually led to significant cost and programme overruns. The architect and the engineers had been getting along very well with communications between the respective teams tending to be informal in nature. However, following the establishment of a co-located design team in Sydney, the relationship deteriorated, apparently because one of the architect's team regarded frequent requests for information from the engineer as an attempt to undermine their office. Eventually, the architect ordered that the door between the two offices be bricked up and gave instructions that engineers would require prior appointments to meet the architects.

Moore and Dainty (1999, 2000, 2001) explored the impact of intra-team boundaries in the efficiency of construction project teams. They found that even where the 'team' had been co-located within a common working space, this had little more than a cosmetic effect with the members choosing to continue to operate within roles demarcated by their professional roles. For construction staff entering the project workgroup later on, becoming an integral part of the decision-making process was difficult, and professional prejudices based on the hierarchy of relationships under the traditional procurement system had reinforced their exclusion. In effect, the team comprised a series of strategic alliances which were bounded by the professional and cultural prejudices of their members rather than any semblance of a working organisational structure. Perhaps the most worrying aspect of Moore and Dainty's findings, however, was that technological solutions had tended to be found to problems which occurred without the team addressing the intra-organisational issues that had caused them. This reflects an emphasis on reactive problem solving as opposed to problem avoidance. Many of these occurrences related to issues concerning the 'buildability' of the project, where a lack of construction management input had led to a failure in foreseeing the impact of impracticable designs. Designers were often required to deal with construction issues, but construction management staff were often excluded from design decisions. Thus, despite the concurrency of the design and construction phases, the transition from design to construction was seen as being akin to process under a traditionally procured contract. If such boundary problems are apparent within integrated teams, it is little wonder that even more severe breakdowns in communication can occur under other traditional procurement regimes.

A further level of complexity is provided by the fact that many individuals who are also members of groups, reside *outside* of the workplace or project. For example, an individual may be a marginal member of a work-based group, but an established member of a trade union with responsibility for maintaining relations with many people on the project. Even where people are brought together to combine their various expertise and approaches in pursuance of innovative and creative solutions (known as communities of practice), this does not necessarily infer that they will develop a common understanding among the group members. Wenger (1998) discusses ownership of meaning in the context of communities of practice (typified by a combination of mutual engagement, joint enterprise and a shared repertoire) wherein it refers to the degree of ability to make use of, effect, or control the meanings that are negotiated by and within communities. Whereas teams do not automatically become 'communities', they generally exhibit at least two of the features of communities of practice – mutual engagement and joint enterprise. Communication within communities of practices will be further explored later in this book given their central importance to knowledge-sharing and problem-solving in project-based organisations.

Of course, most of those who work in the industry assume that discontinuities at functional boundaries will not occur because of the diligence and professionalism of those involved, and because of the well-defined roles and responsibilities which are grounded in the historical structure of the industry. This, it is assumed, is a key benefit of functional specialisation. However, this presents problems for marginal members who have not yet been socialised or inducted to the point where they can be trusted as integral members of the team. Moreover, those with a marginal or temporary involvement to the project, or who operate in an unusual or disparate functional specialism within the supply chain, are unlikely to be accepted or integrated within the project team set-up. Rather than assuming that shared meanings have been established, the sender of the communication must complete the communication model (see Figure 3.3) by provoking feedback to establish the decoding used by the recipient. This is why part of the socialisation of marginal or temporary team members should aim to convey the true meanings of a variety of terms.

The phenomenon and effect of 'groupthink' on team performance

Sociologists have long appreciated that predicting behaviour becomes more problematic when we consider groups and organisations rather than individuals. This is because as more people are involved in a group, the nature of the surrounding communication process must respond to this increasing complexity. One phenomenon which tends to affect the ways in which groups and teams of individuals work together is termed 'groupthink', the tendency of groups of people to be swept along by a particular decision or course of action, rather than use their collective abilities and different perspectives to explore alternative approaches (Katz and Kahn, 1978).

The characteristics of groupthink can be considered in the context of the differences between good and bad communication, with the short perspective being that groupthink is allied to bad communication while its counterpart, 'teamthink', is allied to good communication. Whereas this perspective is not wrong, it is somewhat superficial and a more detailed consideration provides a clearer understanding of the processes at work, thereby underpinning the effective use of communication so as to result in the creation of teamthink rather than the less productive groupthink. Such a deeper examination of this area requires the use of two models: the team lifecycle model and the standard communication model.

The team lifecycle model suggests that a dip in team performance occurs during the entrenchment phase. This dip results from the negative effects of individuals bringing to the team a range of previous values and ideas, not all of which will be positive. Furthermore, there will generally be a wide range of ideas and values represented by the collected individuals from whom a team is to be built, and this diversity represents both a strength and a weakness for teams. It is a strength in that it is a knowledge resource that is essentially synergistic through its value being greater than the sum of the value of each individual's values and ideas. However, in order to achieve this synergistic state the team has to pass through the entrenchment phase or it will remain a group and continue to practice groupthink, thereby not being able to operate at the higher level of performance. It is fair to say, therefore, that the essence of teambuilding *is* good communication and the project manager should be aware of this when attempting to build a new team. There should be considerable support given to the prospective team members as they go through the entrenchment phase in particular. While all projects are battling against the programme of work, the project manager should resist temptation and pressures from elsewhere to 'push' the team through the entrenchment phase of their life cycle. Such resistance can be validated on the basis that any time apparently saved by forcing a team through this phase will almost certainly be lost later in the project due to low levels of creativity and defective decision-making resulting from the presence of groupthink.

Within the context of teambuilding it is essential that all involved are willing and able to both listen and talk effectively and honestly. An important characteristic of groupthink is that the group members are unable to express what may be viewed as non-conformance. Moore (2001) has suggested that it may be more useful to refer to groups as tribes in that tribes, as with 'groupthinkers', place considerable emphasis on conformance to the tribal values and beliefs. Such an approach can be argued to be beneficial in terms of increasing the tribe's probability of long-term survival, but project teams should not be focused on such an issue: projects, as with teams, have their own life cycle and planned duration, and there should be no attempt to prolong the life of the group (an unwillingness to accept mortality is a characteristic of groupthink) beyond that. Along with this, there should be recognition that the team has to focus on the solving of immediate to short-term problems: teams are essentially problem-solving tools and this is where their accumulated knowledge and experiences become a valuable resource. In order to access this resource, the team must pass through the entrenchment phase and through the resolution phase also as it is within this phase that disagreements are resolved. Groupthink does not allow for the communication processes underpinning resolution to take place. This is due to the bad communication resulting from an emphasis on conformance to a clearly defined (and usually narrow in scope) set of beliefs concerning the project and a related set of values dealing with how 'others' (non-group members are regarded as being out-groups who typically pose a threat and therefore should be dealt with aggressively) should be responded to. Groupthink communication is therefore almost entirely about sending a message repeatedly and only responding positively when the 'correct' message is received as a reply. In this manner group conformance (and a low level of performance) is assured.

The potential implications of groupthink to effective communication within the industry are serious and multifarious. As has been discussed earlier, construction is a collectivist activity demanding a range of function specialists to combine their knowledge synergistically in order to develop solutions to complex problems. Innovative solutions to complex problems demand that a range of options are considered, combined and refined. The existence of groupthink within project teams can therefore render the identification and development of creative solutions problematic, reducing the innovative capacity of the group and encouraging linearity in decision-making.

Avoiding groupthink not only demands an effective communication strategy, but also requires leadership in which those with hierarchical authority avoid imposing their ideas, solutions or approaches in favour of a team-based approach. Indeed, achieving teamthink requires all those involved to work through a potentially difficult process. Two factors that are significant in successfully working through that process are the quality of leadership and the level of maturity amongst team members. Much has been written about the whys and wherefores of leadership; what is good leadership, can it be taught, and so on. Although there is no intention to reprise the leadership literature in its entirety at this point, it is worth noting that one common factor that all the leadership models contain is the presence of good communication. Leaders need to be able to communicate (speak and listen) with those who they are leading, as opposed to dictators who only need to instruct and impose: teams will decide whether or not they will follow, while groups will follow without considering the alternative. The project manager must therefore decide whether he/she wish to lead a group, and accept their lower level of performance, or manage a team while being quite clear concerning the difference between the two.

The entrenchment and resolution phase are the key phases of team development in terms of determining whether a project is to be resourced by teams or by groups, primarily because it is not possible to impose team behaviour on a collective of individuals. Attempts to impose on such a collective will invariably result in some individuals deciding that they do not wish to work in such an environment and heading off to new pastures. Those that remain will, at best, perform at the groupthink level. At worst, they will splinter into a number of smaller groups, each of which will become focused on their own interests and create 'mirage' projects (D'Herbemont and Cesar, 1998), and will divert resources away from the project supported by their erstwhile leader. It is therefore vital that the would-be leader uses effective communication (allied with some basic psychology concerning motivation, etc.) as the tool to support the development of the collective of individuals into a team. Part of this development will involve the recognition of maturity in general, and knowledge-based authority in particular, amongst all involved.

Maturity is not simply the fact that someone is older than you, as can be evidenced by asking a simple question: does the individual have x numbers of years experience, or have they had one year's experience x times? In other words, have they learned anything from all their years of being alive, or is life something that was just done to them! Note that the emphasis is not simply on experience of industry; all of an individual's life experiences potentially represent a body of knowledge that could be invaluable. In a groupthink environment the suggestion that someone's experience as a member of their local bird-watching group may provide a solution to a problem encountered by a project would be ridiculed, along with the individual who made the suggestion. This is both poor communication (not listening to the full message) and evidence of a low level of maturity. The individual on the receiving end of the ridicule will be unlikely to make the same mistake twice and in this manner the performance of the group in solving problems is compromised. Project managers should seek to practice good leadership, which will recognise the opportunity for building teamthink, listen to the full content of the message, place it in the context of the sapiential authority of the individual and the needs of the team, allow for a reasoned (and mature) evaluation of the proposal by those team members affected (possibly resulting in it being amended to some extent), and then implement a decision that the team can agree on and thereby choose to follow. The leader has communicated effectively, responded positively to the sapiential authority of others, and given the team the opportunity to follow an action that they have been involved in formulating. All of which takes the team closer to achieving the desired teamthink.

The dangers of oversimplification of communication between group members

An inherent trait when group members attempt to communicate with many others is to simplify the messages they wish to convey. Simplification is no less demanding for the encoder of a message, as refining/reducing messages down to their core components to facilitate understanding and the assimilation of meaning can prove extremely difficult. A good example from popular culture concerns television advertising. For example, consider an advertisement for a new car with a high performance engine. Here, those marketing the product must draw attention to the key advantages of the product rather than its technical complexities if they want to attract new purchasers. Gottschalk (1999) argues that television advertisements are an example of the use of speed (which Virilio (1995) defined in terms of an environment and a social psychological force that transforms what individuals do, how they do it, how they think and feel, and ultimately, what they become) to accelerate the rate at which information is presented. This ensures that individuals have less mental–emotional distance; there is less time to reflect, less delay between interpellation, interpretation, selection and consummation.

Communication in today's postmodernist society is happening increasingly quickly given the onset of ICT (see Chapter 8), with the result that we have less time to spend on coding and decoding messages. The corollary of simplification to enhance understanding in a team or group context can however, be somewhat different. Oversimplification can remove important levels of detail required for appropriate decision-making. Furthermore, it can also increase the risks of misunderstanding on behalf of group members, particularly those outside of the core project group. However, providing too much detail to parties who do not require it can also lead to confusion, misunderstanding and to delayed decision-making. For example, the level of project financial information required by a contractor's project manager about the levels of waste generation on site will be greater than that required by a general foreman with responsibility for managing operatives' waste performance. Although both team members share responsibility for improving the waste management situation, their different situations and roles within the project team demand different levels of information simplification within the communication process. Thus, in this case, the quantity surveyor with responsibility for reporting on waste performance must tailor their report in a way which meets the bespoken needs of the parties involved. Similarly, the project manager and foreman must simplify the information in accordance with the needs of their staff in a way which ensures positive action towards waste reduction. Again, at each stage the potential impact of noise must also be considered, even within fairly cohesive project teams. This example emphasises that individuals within groups and teams have to make careful and reasoned judgements when encoding their desired communication to ensure that fellow members of the group receive the right kinds of information to enable them to act on it effectively. It is important that the ways in which messages are conveyed are tailored towards the needs of the receiver and simplified in accordance with their needs. With the benefit of experience, effective simplification will become instinctive for most people, but it can still be compromised when dealing with unfamiliar people or situations or in times of high pressure with regards to project outcomes.

An effective strategy of team communication demands that the group determine levels of interaction and levels of detail in accordance with the needs of a particular situation. For example, in a crisis such as a fire starting on site, the communication needed would merely be to alert those within the immediate proximity, to raise the alarm and to vacate the site. There would be little need for the precise cause of the fire and situation to be explained in detail to every participant. Conversely, when the cause of the fire is being investigated once it has been brought under control, the communication requirements will increase in detail and complexity in order that the cause can be fully understood. This simple example exemplifies the contingency-based approach used by the majority of project managers, in which approaches are tailored (i.e. they are contingent upon) to the prevailing circumstances (see Loosemore *et al.*, 2003 for an in-depth review of how contingency thinking can be applied to the management of people and communication in construction projects).

Inter-group communication

Up until now, this chapter has largely focused on group communication processes that take place *within* the confines of group or team boundaries. When members interface with individuals, groups or organisations outside of the immediate team environment, however, the communication process becomes even more complex and potentially problematic. The group must find a way of channelling its communication between itself and other groups in such as way as to meet their collective needs, as well as ensuring that it is understood by the other party. Within the construction project environment communication between groups and teams is commonplace, as even within single projects separate teams may exist in dealing with different aspects of the design and production function. Consider, for example, a design team communicating with a project management team on site. Both could be said to be part of the overall project group, but may operate as functionally separate groups. As this book has emphasised, communication across functional and organisational boundaries is fraught with difficulties, and so developing effective communication strategies to facilitate communication between groups is essential to project success. When different members of the design and construction team interact they are effectively creating multiple and parallel communication channels, each of which has the potential to contradict another. Thus, managing inter-group communications is one of the most challenging issues for managers within the construction-project environment.

The team to external environment communication process

As was discussed in Chapter 3, the standard model of communication is one that centres on the processes of encoding and decoding, a process which can act as a significant barrier in the effectiveness of a team to communicate with their external environment. The issue of boundaries was also discussed in Chapter 3, where it was suggested that much of human communication is indiscriminate in that it may not be aimed at a specific individual or group. An example would be a tradesman who stands scratching his head as he seeks a solution to a problem; the action is simply one that has been learned as being appropriate to problem-solving. The fact that it may also communicate that the individual may be confused is largely irrelevant. That is, of course, unless the individual wishes to send a message that will hopefully provoke a helpful response from anyone decoding it. The problem is therefore, not so much one of the ability to communicate, but rather one of communicating in a known-to-be effective manner. These issues apply as much to groups and teams as they do individuals, but have potentially different implications from the perspective of ensuring effective communication. The team to external environment communication process is discussed here in relation to the perspective of the parties alluded to in Figure 3.1:

- *Input* is simply the expression of an intention to communicate with one or other persons. The nature of input is that it disregards unintentional communication such as any body language that the sender is not consciously controlling so as to reinforce the message. *Intention* to communicate makes explicit the need for a specific content which expresses that intention. Content should therefore be appropriate to intention if the communication is to be effective rather than indiscriminate. However, this does not always result in the content being an unambiguous expression; the intention may well be to make the content as ambiguous as possible to anyone other than the intended recipient, as in the case of messages sent using ciphers. This is a conscious decision by the sender. In the case of inter-group communication, the input may be generated from any number of senders.

- *Sender* is the term applied to the person(s) actually intending to communicate. They decide the content to be communicated, for example, 'Concrete is needed in bay 3 immediately', the channel of communication (see later) and also on the manner of encoding to be applied. In the context of intra-team communication, the encoding will most probably be on the basis of a shared perception of reality; everyone in the team is likely to know where bay 3 is, what type of concrete is required, any surface finishes to be applied, and so on. The form of expression given to the content will usually be one that is most capable of being correctly decoded by all the intended recipients. In the case of inter-team communication, it is possible that the shared perception of reality may not already be in place (as with common understanding of a specific technical or non-technical language, agreed objectives to be valued, etc.), and therefore needs to be created. This may require several messages to be communicated prior to the message containing the desired content that needs to be decoded correctly in order for the sender's intention to be understood. These 'prior' communications have the objective of establishing a basis for a shared perception of reality, and may be regarded as developing a means of translation between the various team-specific languages that may be involved. This process may use more than one channel.

- *Channels* are also referred to as the medium of communication: the material through which the content is sent. Several materials can be identified. Acoustic communication, such as speech, uses the channel provided by the material of air. Visual communication, such as sign language, uses the channel provided by light. Written communication, such as letters or books, traditionally uses the channel provided by paper, but also now uses the channel provided by computer monitors, etc. Other channels can be identified, but in all cases the channel must be appropriate to the content if communication is to be effective. Someone who is completely blind, for example, is not going to find a visual communication channel an effective means of receiving content. Such a situation could be regarded as an example of noise.

- *Noise* refers to anything that interferes with the satisfactory decoding of content, and is not confined to, for example, loud music that drowns out verbally communicated content. A coffee cup placed on a letter can result in noise through the creation of a stain obliterating some of the written content. Communication effectiveness is therefore reduced as the receiver's ability to decode what remains will be reduced. The extent of the noise in this example will depend on how important the obliterated content was to the encoded message, and also on ability to decode an impaired content on the part of the receiver. The more complex and multifarious the team communication processes, the more likely that noise will impact negatively on the efficacy of the communication process. In inter-team communication processes, any individual member can contribute to the noise which will collectively impede the efficacy of the process.

- *Receiver* is the term applied to the person who decodes the content as received. This decoding can be in three stages. The first stage is recognition of the existence of 'content', followed by the application of a relevant decoding technique. Finally, the decoded content is placed in a suitable context. This latter stage can also be regarded as a process of translation. For example, the receipt of a fax provides evidence of a clearly bounded channel; the message must be contained within the page(s) of the fax. The decoding process is already underway, and the second stage is one where the receiver selects a decoding technique to apply to the content. The letters of the alphabet may be separated from numbers, and then individual letters decoded into words, and words into sentences. This is where translation is applied so as to place content into a meaningful context and thus achieve output. Again, in an inter-team communication context, there may be multiple receivers all decoding the information in different ways, thereby increasing the likelihood of miscoded information.
- *Output* is simply the content as decoded by the receiver, with the trick being to make certain that input and output are sufficiently similar. For this to be achieved, the code used by the sender must be one that is also available to the receiver; for example, sending content encoded in French will not provide effective communication if the receiver can only decode in German.

Within models stemming from Shannon and Weaver's (1949) original theory, if the encoding or decoding is defective then the effectiveness of communication will be diminished. In the worst case, there will be no effective communication at all, with either the total inability of the recipient(s) to decode the message or a completely inaccurate decoding of the message being achieved (which often has more serious implications). The process of encoding/decoding and the potential for problems occurring was discussed in some detail in Chapter 3, but is worth returning to in relation to the way in which teams communicate with their external environment (i.e. parties outside of the immediate workgroup). In the standard communication models, such interactions comprise linear communication between a series of individuals.

For example, a member of the construction management team may request a clarification in a design detail from a member of the design team. The designer may then explore the implications with the quantity surveyor, who may in turn consult the specialist contractor as to the cost implications of the revised details. This 'chain' of communication occurring in response to the initial communication by the member of the construction management team is an example of what Walker (2002) defined as 'sequential interdependence'; a situation where one part of the process depends on others for its completion. However, in reality, the organisation of construction processes is such that the matrix structures operated by most construction companies (see Loosemore *et al.*, 2003) result in a potentially highly complex series of relationships in which a team member may be involved in several communication processes across several projects simultaneously. Considering that each project may operate on a different contractual/procurement basis (e.g. D&B, management contract, traditional procurement, etc.), a different set of relationships may be generated within each project, even if the same parties are involved (which is in itself unlikely). The nature of the communication environment will therefore differ for each project undertaken, which places considerable demands upon the project participants who must tailor their approaches to the particular needs of the project and the teams/individuals with whom they must interact and communicate.

In his perspective on sequential interdependence, Walker (2002) is essentially considering completion as being achieved by the addition of varying contributions throughout a linear process of construction. For examples, external walls being completed by the linear contributions of excavation, foundations, brickwork, blockwork, windows and doors represents a linear and sequential process of construction. In this context, there is a deliberate and planned intention to achieve completion through changing the product at each stage of the process. However, Walker notes that this perception of the construction process is overly simplistic and that it is arguably more accurate to regard construction as representing a number of *reciprocally* interdependent processes. Reciprocal interdependency infers a relationship based on the iterative development of the 'product' by all appropriate players. Unfortunately, successful management of reciprocal interdependency requires higher levels of skill and effort than does sequential interdependency (Walker, 2002), which suggests that traditional (linear) models of communication may not apply.

In order to put the preceding discussion into context and illustrate the complexity of communication between teams and external environments, Figure 5.1 presents a model of communication between a single sender and two teams, each with three members. For the sake of simplification the sender is assumed to be an individual rather than an organisation, and the feedback channel has been omitted for clarity. Each member is represented by a series of boxes. Box P1 represents that member's primary role which, in

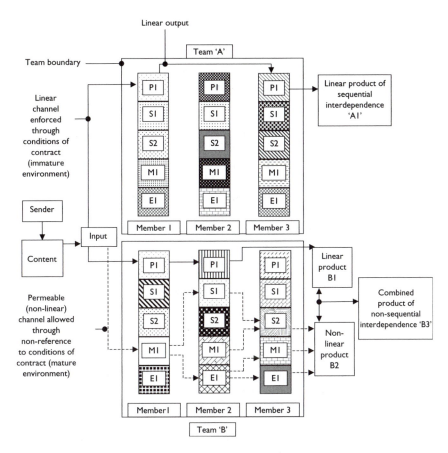

Figure 5.1 Linear and non-linear channels and outputs in teams.

the linearity perspective, would be his/her functional specialism (e.g. architect, quantity surveyor, project manager, etc.) P1 has the same meaning in the non-linear model, but this model also includes other functions that the linear model does not consider. Thus when a team member receives 'content' from the sender, the non-linear model acknowledges that along with his/her functional specialism he/she brings other functions to the decoding process. S1 and S2, for example, are their secondary roles (an individual may have several or only one secondary role (see Handy, 1999), although all team members in Figure 5.1 are shown as having two for the sake of consistency and in this example no two members have the same mix of secondary roles). The differing hatchings represent differing secondary roles.

A transformational view of organisations is one that is essentially non-linear (Banner and Gagne, 1995) and argues that individuals, along with their functional expertise, possess varying levels of maturity (the perception that they are persons who can be trusted, possibly even regarded as an 'established' member of a society or community), M1. They also possess a level of what may be referred to as 'expertise', E1, which differs from that related to their functional specialism in that it is a general recognition of how life works; they know the rules of the game and how best to play it. Within the linear model, both of these functions are regarded as simply adding noise that is additional to that regarded as arising from the secondary roles, and the sender may well try to prevent the receiver bringing these functions into use during the translation process by seeking appropriately linear channels. In Figure 5.1, for example, the product of Team 'A' arises solely from the principle roles of two team members. Although in some instances this degree of control over the communication process may be vital, it can also be argued that in other instances such control results in an immature working environment and an impoverished product. In imposing such rigid control, the sender is attempting to place a largely impermeable boundary around the workings of the team, thus creating a method of working which results in the team behaving as a closed system having limited ability to self-regulate. The team therefore becomes dependant upon the sender for the content that allows them to respond in the event of any change in their external environment. This presents risks in that the response time of the team slows down to that at which the sender is able, or willing, to supply the required content, plus the time taken by the team to achieve translation of that content. In comparison, Team 'B' in Figure 5.1 have been allowed to access their secondary, experiential and maturity roles in addition to their primary roles and to do so in a non-sequential manner. Thus team member 3, for example, is 'allowed' to contribute without any reference to their primary role. Such a method of contributing to the product can only take place in a relatively mature working environment, with high levels of recognition by team members of their co-members' abilities in a range of roles. This mature environment produces both linear and non-linear products, which can in turn be combined if desired to produce a 'richer' product than is possible in the immature environment imposed on Team 'A'.

Each of the secondary roles can be argued to represent 'noise' in the sense of the linear model of communication in that the content was encoded on the basis of being decoded in the context of the receiver's primary role. The greater the number of secondary roles a member is capable of adopting, the greater the potential for noise in that the sender is not aware of the level of expertise the member possesses in each of his/her secondary roles. The translation process will be affected to an unknown degree. Thus the sender generally attempts, within the linear model, to constrain this immeasurable affect on translation by making use of linear channels. The most extreme example would be content encoded in the context of contractual responsibilities in an attempt to keep the receiver within their primary role, as it is that expertise that is being paid for. Such messages will typically have a highly 'official' tone (possibly using references to specific clauses in the contract) and be full of detail that is present largely as a mechanism to ring-fence the translation process. However, team members are only human and will inevitably bring with them more than their primary and secondary roles.

To put the process illustrated in Figure 5.1 into a more practical context, it is useful to examine the ways in which it can impede communication within a practical situation. As was discussed earlier, Moore and Dainty (2000) found that even within supposedly integrated D&B construction project teams, problems of communication arose. The majority of the problems identified concerned the flow of information, and how this was being constrained by some members of the team through the selection of a channel that forced other team members to act as a largely closed system when dealing with certain aspects of the project. The intention of forming integrated teams is arguably to increase the openness of the project system by removing boundaries around specialisms. Although it may not be an explicit objective when forming an integrated team, the increased openness that should result provides increased opportunities for non-linear communication as the content that arrives as input flows freely around the team environment. Increased openness allows team members to direct content to other members on the basis of not only their primary role(s), but also their secondary role(s), expertise role(s), and maturity role(s). The theoretical benefit of this non-linear mode of communication is the potential to increase creativity in problem-solving.

In the case of the D&B integrated project team investigated by Moore and Dainty (ibid.), some of the architect members refused to communicate directly with the construction manager members. Instead, they formed a channel that went via the quantity surveyor members, thus imposing an unnecessary additional translation on the content. However, they appeared to be willing to accept the risk of some mistranslation of the content as the addition of the quantity surveyors to the channel achieved the objective of placing the construction manager members in something close to their position in the hierarchy of a traditional procurement route. The architect is, of course, at the top of the hierarchy in such a procurement route! The D&B project team can be argued to have missed an opportunity to realise at least some of the potential of a non-linear approach to communication. This may be due to the need to adopt a transformational way of thinking in order to achieve this potential, and the difficulty of replacing years of training in traditional, linear and transactional ways of working with the very different transformational way of working. This affords the opportunity to increase the richness of content being communicated by allowing team members to operate in a manner that is arguably more natural, rather than imposing the artificial channels defined by the linear model. The full diversity of the project team, in terms of all the different roles that represent each individual, begins to become available to both the sender(s) and the receiver(s), and content can be coded in ways that would not even be considered by the linearity model.

A good example of transformational communication in construction is provided by the architect Daniel Liebskind and the way in which he codes his communication about the buildings he is designing (see Muller, 1997). His competition entries in particular are strikingly different from the traditional submissions that are largely based on a linear model of communication. Liebskind's submissions are more non-linear than the norm, as evidenced by, at the first attempt at decoding, their apparent lack of any comment about a building. However, they are arguably specifically not aimed at the receiver's primary role, and it may be more appropriate to think of them as being aimed at the maturity role (as Figure 5.1), and possibly also the expertise role of the receiver(s). The difficulty appears to be that the receiver has been trained (as part of becoming an established member of their group) to apply his/her primary role (architect, client, etc.) when seeking to decode the content received.

Barriers and obstacles in the team to external environment communication process

Considering the complexity of communication between parties and teams involved in the construction process, along with the extent of interdependence in the construction process, it is unsurprising that effective communication is often impeded. Even ICT may not always enable effective communication as they can inhibit explanation of meaning between project participants (see Chapter 8). Project environments typically involve teams and individuals that need to observe the dynamics of the different levels within the environment hierarchy (project environment, activity environment, task environment, etc.) It is therefore important to understand the impact that these dynamics have in both inhibiting and creating opportunities for improving communication between groups and teams.

A good example of how obstacles emerge in construction communication processes is provided by Brown (2001), who explored communication in the design process. Brown identified a deficiency that arises from construction players persisting in the application of an inappropriate model of communication known as the 'service gap'. This theoretical perspective was originally suggested as existing in five forms which Brown (2001) summarises as:

- *Gap 1* – between consumer expectation and management's perception of consumer expectation.
- *Gap 2* – between management's perception of consumer expectation and management translations of them into quality service specifications.

- *Gap 3* – between service quality specifications and the actual service delivery.
- *Gap 4* – between actual service delivery and external communications about that service.
- *Gap 5* – between actual service delivery and the consumer's perception of the service.

Although each of these gaps indicates individual relationships that combine to form a network of reciprocal interdependency, there is also a communication issue identified by them, and which Brown refers to as gaps between expectation and perception. These gaps not only exist with regard to the service issue, but also the product. Recognition of the existence of these gaps is an important step in the development of an appropriate model of communication for construction, but the fact that communication problems persist nearly 20 years after the identification of them suggests that there may be other issues at work. One possible issue is that of the gap between expectation and *cognition*. The perception (awareness or observation) of a message actually precedes cognition (knowing or conceiving) in the communication process. On this basis, the manner in which language is used, particularly where there are a number of commonly accepted meanings for a given term, could well contribute to the service and product gaps. A simple example is that an individual player within a group may well have a different cognition with regard to a concept that is being communicated than that of the other members of the group. However, an important step in bridging the service gap must be to facilitate the reaching of a common cognition of reality as a follow-on from a common perception of that reality. By referring again to the reciprocal interdependency model, this suggests a situation where each interdependent relationship represents an opportunity for both a service gap and achieving of a common cognition of reality. If the latter can be achieved, then the former should be obviated. The achievement of a cognition of reality that is common to all players involved in a particular interdependent relationship is therefore, arguably, the prime objective of good communication in construction. Achieving this will, however, mean that each of the players must exhibit sufficient maturity to modify their cognition when they perceive new evidence that confronts their existing cognition, rather than seeking to defend a consistent, but inappropriate, cognition.

In the context of construction communication, the intention often appears to be the reverse of this; 'product' completion is achieved only if the product goes through the various stages of the communication process without changing. In much of the day-to-day communication involved in construction projects, the intention is to pass a concept from one individual or group to another without it undergoing change. In other words, such messages must remain constant, irrespective of any differences in language or culture (or the extent of any other forms of noise) between any of the contributors who are deemed to be 'sequentially interdependent'. It can be argued that this perspective (of constancy regarding the message) fits reasonably well with the 'production' emphasis in Walker's view of sequential interdependence. There is, however, also the perspective that communication can be part of a creative process within which it is used to fuel the evolution of ideas, such as in the development of design concepts and in problem-solving. In such cases, completion can only be achieved successfully if there is the potential for each of the contributions to bring about change in the 'product' being communicated. To add to the complexity of managing such a communication process, the type of interdependence in 'creative' communication may not be confined to being sequential, in that the product may be moved backwards as well as forwards throughout the sequence of contributors as design tends to be an iterative process. This form of reciprocal communication can be argued to be less linear than the form within truly sequential interdependence and a good project manager will seek to be aware of this difference when managing the flow of communication within the project environment.

On the basis of the preceding discussion, it can be seen that the nature of interdependence in construction results in a potentially complex matrix of relationships in which a team member may be involved in several projects simultaneously. Considering that each project may operate on a different contractual basis (a different set of relationships may be generated within each project), the nature of the relationship (in communication terms) between the contractor and the client, for example, may be heavily constrained through their defining of their relationship by reference to the form of contract used. Standard forms that are highly detailed may well encourage a relationship based on identification of responsibilities throughout the project lifecycle. Such a relationship serves to strengthen the perception of differences (rather than cognition of the realities of the 'production' environment) between the parties involved and may, therefore, hinder the communication process. In extreme cases it is possible that communication is only possible on the basis of a mutual understanding of the contract terms, and when such mutual understanding is not achieved the relationship becomes further complicated as both parties add further contributors to the sequence of communication (e.g. in the form of legal advisers). Such conflict-oriented approaches to communication merely serve to strengthen the barriers (and widen the service gaps) that already exist between the different parties involved in the construction process and a good project manager will seek to avoid resorting to such actions, irrespective of how uncommunicative other parties within the project may choose to be. Although each of these parties will typically be able to communicate effectively within their team or group boundaries (as in intra-team communication) on the basis of agreed meaning of language (common cognition), problems start to emerge as the sequence of communication moves out of the team environment. The success of team to team (inter-team) communication will depend upon how far removed each team is from a mutual understanding of language. For example, a bricklayer may know what is meant when referring to a 'quoin', but do members of the plumbing team?

As was emphasised in Chapter 2, the nature of the environment in which communication takes place may be different both between projects and within each project. However, although sequential interdependence (along with other less linear forms of interdependence) has an impact on the communication environment for teams, it can be argued to have an essentially production-oriented emphasis. Thus, it may not be as relevant in this context of the role of the professional team as is a concept such as ownership of meaning. Wenger (1998) discussed ownership of meaning in the context of communities of practice (typified by a combination of mutual engagement, joint enterprise and a shared repertoire) wherein it refers to the degree of ability to make use of, effect, or control the meanings that are negotiated by and within communities. Although it is not being suggested that teams are automatically the same thing as communities of practice, they do generally exhibit at least two of the features of communities of practice (mutual engagement and joint enterprise, with a shared repertoire being a possibility for certain teams) and are therefore worthy of consideration in the context of how ownership of meaning affects communication. This theme will be further developed in Chapter 8 in relation to ICT.

Summary

This chapter has raised and elaborated a number of challenges, both for managers in the construction industry and to some of the established approaches to communication between and within teams. It has suggested that the essence of teambuilding *is* good communication and the project manager should be aware of this when attempting to build a new team. However, at an intra-team level, role-based boundaries and patterns of interaction define a problematic context for ensuring effective communication within them. At an inter-team level the patterns of communication become far more complex and the potential increase for noise. The most significant of these problems lies in the need to understand the different approaches to communication that flow from the linear and non-linear models. When considering these models, the reader should be aware of the impact of the linear model on their own approach to the decoding of content. The problem of ownership of meaning was raised in the context of levels of trust and how these differ between established and marginal members of construction teams. Established team members should be aware of how they have been socialised into a series of shared meanings by their interactions within the 'team', and the impact of this on the decoding process.

Critical discussion questions

1   Map out a diagrammatic representation of the communication structure of a simple project. Identify the potential structural cultural barriers both within and outside of the team which have the potential to impede the efficacy of the communication process and devise a strategy which can overcome these obstacles to effective project performance.

2   Considering the numerous constraints on team and group communication outlined within this chapter, it is obvious that a fresh approach to addressing communication problems is now required if project participants are to break down traditional barriers to understanding and information flow. However, changing the ingrained cultures which have become part and parcel of the industry's working practices will arguably demand root and branch reform of both the development of people who work within the sector and, just as importantly, the ways in which parties involved in construction processes work together and interrelate. Discuss the ways in which such change could be brought about in the construction industry of the future.

Case study: developing high performance teams: soft and hard communications initiatives at the Simons Group

Introduction

Simons Group is a privately owned Construction, Property Development and Consultancy Group with an annual turnover in excess of £250 million. It is based in Lincoln, and has offices throughout the East Midlands, South East, North East and North West England. The company has three principal areas of business, Construction, Property and Consultancy services, and includes design/architecture, environmental, mechanical and electrical design and project management/cost consultancy functions (see Figure 5.2). Thus, Simons Group aim to provide a 'one-stop-shop' for clients and consultant advisors wishing to procure construction services. This case study examines how this leading company has improved communications throughout its business by developing a culture of continuous improvement. It focuses on the company's adoption of the Belbin Team-Role Self-Perception Inventory toolkit, which it has combined with other ICT tools to improve its team working and communication practices.

Improving project team performance

During the late 1990s, senior managers at Simons Group were questioning the manner by which its project staff communicated with other project

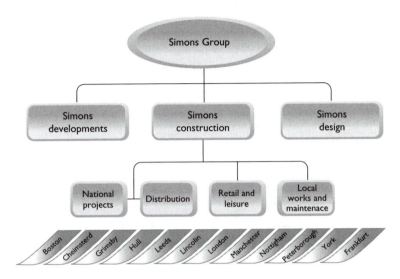

Figure 5.2 Simons Group structure.

stakeholders, both up and down the supply chain. In common with many other construction organisations that were seeking closer collaboration, the company set about revitalising their relationships with clients' contractors and suppliers. In order to achieve this, it became apparent that there was a need for the Simons Group project staff to understand how they interacted with colleagues and project personnel employed by other companies. During in-house meetings, several project personnel expressed concerns regarding the performance of project 'teams' and it was concluded that superior performance was not always achieved because the inter-organisational and interdisciplinary project personnel were in fact performing as a work group. The synergy expected from construction project teams did not always materialise and this had led to difficulties in providing clients with superior performance. Typically, many of the problem incidents that had occurred in projects were attributable to behavioural traits and project communications were considered to be key issues where measurable improvements could be made.

The first stage of the initiative involved understanding the behavioural traits of Simons Group Construction division project staff. It was necessary for the organisation to commit itself to a form of reflective analysis before embarking on an improvement programme to change the operating culture of projects in which the company was involved. The Human Resource (HR) department at Simons Group sought external assistance with this exercise and employed a consultant with expertise in psychological profiling. Having assessed various models and toolkits, the consultant recommended the use of the Belbin Team-Role Self-Perception Inventory toolkit. The use of the toolkit was piloted with several groups of employees from the Construction Division prior to being rolled out throughout the organisation. During this period, the HR team were also collaborating with senior staff in the Information Technology (IT) department. The IT staff had been developing a collaborative project portal that would enhance communications during projects and it became clear that a dual-faceted approach aimed at improving both 'hard' systems and 'soft' behavioural aspects of communication would enhance the likelihood of each initiative being successful.

Improving team performance through the Belbin Team-Role Inventory

During a period of over nine years, Meredith Belbin and his team of researchers studied the behaviour of managers from all over the world. Managers taking part in the study were given a range of psychometric tests and put into teams of varying composition, while they were engaged in complex management exercises. Their different core personality traits, intellectual styles and behaviours were assessed during the exercise. As time progressed different clusters of behaviours were identified as underlying the success of the teams. These successful clusters of behaviours (or team-role types) were then given names, which are summarised in Table 5.2. The accurate delineation of these team roles is critical in understanding the dynamics of any management or work team.

Project case study

An example of the approach adopted by Simons is within a strategic partnering relationship that they developed with a major client, who provide residential accommodation for patients with learning difficulties and behavioural disorders. To date they have worked on four main projects, with a further two programmed. The projects have been procured using the project-partnering contract – PPC2000, which in itself encourages more transparent relationships between project stakeholders.

All key Simons' staff involved in these projects undertook the Belbin Team-Role Self-Perception Inventory, together with the client, the consultants and the key suppliers. The assessments were compared for compatibility and expected performance. The report produced indicated an even spread of expertise and team-role types and so in theory, the areas of conflict could be removed and any shortfall in expertise filled. However, due to programme pressures, no changes were made to the team composition. The Core Group was made aware of the potential areas of conflict. Within the site team and the project team, further assessments were carried out with, for example,

Table 5.2 Belbin team roles

Belbin team-role type	Visual representation	Contributions	Allowable weaknesses
Plant	PL	Creative, imaginative, unorthodox. Solves difficult problems	Ignores incidentals. Too pre-occupied to communicate effectively
Co-coordinator	CO	Mature, confident, a good chairperson. Clarifies goals, promotes decision-making, delegates well	Can often be seen as manipulative. Off loads personal work
Monitor evaluator	ME	Sober, strategic and discerning. Sees all options. Judges accurately	Lacks drive and ability to inspire others
Implementer	IMP	Disciplined, reliable, conservative and efficient. Turns ideas into practical actions	Somewhat inflexible. Slow to respond to new possibilities
Completer finisher	CF	Painstaking, conscientious, anxious. Searches out errors and omissions. Delivers on time	Inclined to worry unduly. Reluctant to delegate
Resource investigator	RI	Extrovert, enthusiastic, communicative. Explores opportunities. Develops contacts	Over – optimistic. Loses interest once initial enthusiasm has passed
Shaper	SH	Challenging, dynamic, thrives on pressure. The drive and courage to overcome obstacles	Prone to provocation. Offends people's feelings
Team worker	TW	Cooperative, mild, perceptive and diplomatic. Listens, builds, averts friction	Indecisive in crunch situations
Specialist	SP	Single-minded, self-starting, dedicated. Provides knowledge and skills in rare supply	Contributes only on a narrow front. Dwells on technicalities

Source: www.belbin.com/belbin-team-roles.htm (accessed 05/08/05).

the surveying team consisting of the Site Project Surveyor, Clients' QS & Key Account Manager. The assessments created a great deal of interest, particularly among the senior members of the team. In some instances there was disagreement with the findings, but this in itself was positive in that it generated discussion around their strengths and contributions. In later years the team attended workshops conveyed by an external facilitator and they analysed potential areas of conflict, that is, one team member was thought to respond negatively to criticism. The team discussed ways of managing the various personalities without conflict or breakdown.

The initial project was considered to be very successful and the team continued to work very closely together (they have subsequently completed three more schemes). Relationships have been strengthened and the participants envisage maintaining the relationships on future schemes. Team communications have been excellent throughout and have been substantially enhanced by using the Simons web-portal system, which is linked to all team members including the key members of the supply chain. As a means to accommodate any team replacements or new members in the future, the Belbin profiles library is periodically updated in order that an accurate and up to date database is maintained to facilitate future team development activities.

Supporting technologies for promoting effective teamwork

Simons are firm believers in the sharing of information throughout their projects and to this end they have developed a collaborative project portal (simonstogether.com) that has been developed in a way that unifies the virtual team's information, facilitates the project management process and results in a significant reduction of time lost through ineffective communications. Experience of web portals has shown that access speed is a critical success factor. Without speed, the team could soon revert to paper, email or fax, losing the cohesive benefit that a central searchable knowledge repository would give. This portal is different in that it can be used not only connected to the Internet, but also offline. Offline working improves performance by up to 700 per cent and allows input when network connection is impractical. Another feature is that each project portal has a fax number that enables direct contact with subcontractors without web access. The system is capable of being used throughout the project by all parties, thereby capturing a complete record of the experience through every piece of written or drawn communication. All information stored in the searchable database may eventually be updated and turned to CD for issue to the project team, the building owner or suppliers, giving them a complete set of 'as built' project records. The knowledge gathered may be reviewed by the team and if appropriate reused on subsequent projects. Workflows become more formalised and automated with the passing of information such as drawing comments, approvals and instructions quickening the process and ensuring uniform handling.

The benefits derived from using the web portal have included the elimination of the risk of losing important files and documents, an improvement in team communication and the coordination of tasks and activities. The improvement in communications is realised by the project team operating in a proactive manner and this has led to a reduction of errors and rework in the building process. Furthermore, the system has led to savings in time during the comment and approval process. A further benefit associated with the transparency promoted by the system has been a reduction in disputes in Simons' projects.

Conclusions and lessons learned

Simons' proactive approach to ensuring communication from their project teams and integration of their project delivery processes has paid dividends in terms of supporting their ability to deliver more effective solutions to their client base. Combining the benefits derived from applying the Belbin test (i.e. increased awareness of team roles and behaviour) with their web-based knowledge management portal, has enabled them to support the formation and development of more synergistic teams. Moreover, it has enabled them to support the development of a corporate culture that values people by encouraging well-motivated employees to commit themselves to company success and thus, to a high level of quality and customer service. In parallel with these initiatives, Simons have established the '50:50 Vision programme' which is aimed at creating a more diverse workforce and hence, at strengthening their customer focus through improved internal and external communications. The communications networks and improved working that have flowed from this initiative have further enabled the company to improve its team and group communications throughout the business. This case study has provided an excellent example of how a company can combine 'soft' teambuilding initiatives with 'hard' technological solutions to propagate a high performance team environment.

Chapter 6

Organisational communication

Introduction

Communication can be regarded as the 'substance' of everyday organisational life (Eisenberg and Goodall, 1993: 18). Although information technology can facilitate communication (see Chapter 8), effective information flow is ultimately achieved through effective interaction amongst people who operate within the context of their groups, teams and networks. Such interaction is neither random nor accidental, but is contained within structures or organisations which, *inter alia*, channel communications between people (Thomason, 1988: 400). Accordingly, this chapter explores organisational communication in terms of the ways in which people communicate within and across the boundaries of the firm. This will inevitably involve information flow across professional, departmental, team and/or functional boundaries and so it accounts for some extremely complex processes of interaction, especially within larger and more complex organisations. As was explored in Chapters 2 and 5, in order to survive and respond to the demands of the industry in which they operate, construction organisations tend to be highly complex and dynamic in nature, as well as being highly responsive to the change which inevitably envelops the projects and the supply chains from which they are constituted. Accordingly, in this chapter the nature of communication within and across construction organisations is explored in order that the obstacles to effective communication flow can be identified and subsequently overcome by those responsible for the success of project outcomes.

The organisation as an arena for communication

The study of people in organisations is one of the most established lines of research enquiry. Over 60 years ago, Chester Barnard published his seminal work on *The Functions of the Executive* where he suggested that 'In the exhaustive theory of organization, communication would occupy a central place, because the structure, extensiveness, and scope of organizations are almost entirely determined by communication techniques' (Barnard, 1938).

In more recent years, this realisation has resulted in theorists having begun to think more systematically about people's roles and behaviour within them (Handy, 1993: 20). However, although communication underpins the majority of organisational theory (particularly in terms of delivering change as will be explored later in this book), it is one of the least recognised aspects of management by organisational scientists and practitioners (Church, 1996). Reasons for this lack of understanding remain speculative, but it may be because social communicative exchange and the patterns of interaction that occur within organisations are often taken for granted, despite the advantages that understanding an organisation's communication dynamics can provide. Indeed, most organisational research has been oriented towards meeting the needs of management, rather than understanding the functioning of the organisation itself (Pacanowsky and O'Donnell-Trujillo, 1990). This belies the crucial importance of communication (both formal and informal) to the overall functioning and success of organisations and the endeavours with which they are involved.

In recent years there has been an increasing trend towards research which has considered the ways in which people interact and more particularly, the ways in which they communicate. Whereas in the past lines of communication were assumed to be linked to the management hierarchy (and thus fairly linear in nature and related to formal authority), the contemporary view of organisations recognises that information flows in many directions and in different ways between colleagues at all hierarchical levels (consider for example, the discussion in Chapter 5 around the non-linear communication model and Figure 2.1 in which the informality of communication within organisations is demonstrated). In this respect, the cultural context of the firm is key to understanding *how* communication takes place and how effective it is. Thus, understanding a company's culture is fundamental to understanding how communication works within the firm. Indeed, structure is nothing more than the relationship between organisational positions or roles; as relations between people change, so does structure. This is particularly the case in project-based organisations which tend to adopt the team-based 'matrix' organisational structure where functional and project reporting structures often differ (see Loosemore *et al.*, 2003). Nevertheless, no matter how small or organic an organisation is, there will inevitably be a need to make decisions on how communication will be managed and what forms of channels will be used. This recognises that the appropriate management of the communication process can manipulate the quality and appropriateness of information flow. This is not to belie the impact that informal and undefined processes will have on communication, but there is a greater probability that communication will occur as intended if some form of structural governance is put around the communication process.

Internal and external organizational communication contexts

Communication is an essential aspect of the functioning of an organisation as well as in the its's information exchanges with its environment (Rogers and Agarwala–Rogers, 1976: 7). Thus, organisational communication can be considered in either internally or externally defined terms:

- The *internal* dimension focuses on ensuring effective communications between managers and workers in different parts of an organisation, particularly project staff and central HRM departments. This is vital for regulating employee behaviour (in a way that ensures congruence with organisational goals), for innovation (in changing the way things are done), integration (bringing together processes in a way that helps to deliver on the organisation's objectives) and information (passing on the information needed by employees to perform in their jobs).
- The *external* or inter-organisational communication dimension focuses on information exchange with external (outside) parties, such as suppliers, clients, local communities and trades unions. Here, the focus is on the way in which the organisation manages its communication processes with the outside world.

Although these two perspectives to some extent represent different kinds of processes, internal and external communications must be seen as mutually reliant and intertwined, as if one is ineffective, then the other is likely to be detrimentally affected. Consider a construction company with poor communication between its commercial and production managers within its projects. This is likely to lead to further problems in the messages conveyed to the clients, suppliers and subcontractors with whom it interacts, or perhaps even the corporate image of the organisation as a whole (see Chapter 7). No matter how effective internal communication mechanisms are, a failure to communicate effectively with external parties will undermine confidence in the firm. Thus, a failure of either process is likely to impact detrimentally on the organisation's performance. The case study at the end of this chapter presents an excellent example of how internal and external communications can be managed effectively.

Formal and informal organisational communication

A further distinction in the types of organisational communication can be drawn by acknowledging the existence of both formal and informal forms of interaction. Dow (1988) refers to two conceptual schemes of organisational structure; configuration and co-activational. As was discussed earlier, organisational charts represent the configuration view using vertical lines and hierarchical relationships to imply management authority. This is the view taken by construction practitioners who believe that project structures can be formally designed. In contrast, the co-activational view is that structure is *inferred* through regularities in the behaviour of project participants over time. This view was reinforced by Wofford *et al.* (1977) whose examination of informal communication revealed that people communicate because of their own psychology, circumstances and because they want to, not just because the organisation tells them to. It is arguably these informal communication patterns that are more crucial to the efficacy of temporary organisations such as construction projects, as the network of relationships is so fluid and subject to change. Thus, it can be seen that fully understanding organisational communication (and the motivation behind it) demands knowledge of both formal and informal structures that exist within the bounds of the firm. This is discussed in more detail later in this chapter.

Communication as a determinant of organisational structure

As important as organisational structure is in influencing communication, communication can also be shown to affect the way in which organisations develop in a structural sense. Conrath (1973) posits that organisational communication data may provide the 'essential ingredient' for the study of organisations. Roberts and O'Reilly (1978) go a step further by suggesting that organisations could actually be described as complex, overlaid communication networks. Indeed, proponents of this view such as Fisher (1981) suggest that organisational structure and communications are virtually inseparable. What becomes apparent from the literature is that communication helps to define organisations by encouraging the creation of the structures from which organisations are comprised (Weick, 1987). In this respect the communication processes (both formal and informal as alluded to earlier) represent the fabric of the organisation to both employees and external parties. Numerous authors have explored the mutual influence that communication and structure have on each other as will be explored later.

The structure of an organisation can be defined as patterns of interaction rather than in terms of the organisation chart. Dalton's (1959) work revealed that managers engage in extensive interactions beyond the hierarchically specified communication channels and are highly embedded in informal structures of interaction. At a project level, Hopper (1990) suggested that construction projects depend on informal organisational behaviour more than organisational types. Hopper refers to informal structure as a shadow or parallel structure, arguing that it is built around three sets of legitimate needs:

1   The need to maintain links of communication, coordination, problem-solving and decision-making when the established structure isn't working properly.
2   The need to maintain these links when it is working properly in order to interpret, translate and expedite the requirements of the established structure.
3   The need to form informal cabals (support groups) as an efficient way of getting things done, sustaining personal relationships and avoiding the maze of structure.

It can be appreciated that these three sets of mutually supportive needs can all be seen to exist within the involvement climate necessary for successful construction projects. For example, construction projects are known for presenting a plethora of complex problems which must be resolved through interdisciplinary teamwork, which itself demands that people are able to interact without recourse to organisational demarcated boundaries and roles. Thus, the communication of any project will be tailored to suit the interaction requirements required. From such structures, informal communities of practice will emerge as those individuals and teams with differing skills and knowledge sets come together to pool their resources and solve problems together. Thus, it can be appreciated that in many respects the informal patterns of interaction that stem from the social structure of an organisation are both a necessary and defining aspect of organisational life. However, despite this widespread acknowledgement of their dynamic inter-relationship, as Monge and Contractor (1988) point out, networks of social structure (of which communication is one type) are very difficult to identify because they are comprised of abstract human behaviour over time, rather than physical material. It is important, therefore, that project managers recognise and understand the role of communication in shaping the workings of their project and/or organisation if they are to manage the patterns of interaction within them that will ultimately shape project outcomes.

Understanding communication within organisations

Understanding communication within organisations is vital to understanding why they succeed or fail. This is because without effective communications systems and procedures, they cannot manage the complex flow of information necessary for interaction with their internal and external environment. Expressed simply, good communications are required in order to achieve coordinated results (Armstrong, 2001: 807). Accordingly, the study of communication in organisations usually examines the flow of information through channels and networks and the contents of messages sent (Thompson and McHugh, 2002: 260). This is a difficult undertaking as not only will the structures and systems need to be understood, but also the cultures and norms of the organisation and the ways in which they come together to shape information flow. Indeed, organisational structure, as defined by communication patterns, provides the most profound insights into organisational power. Hence, in many regards understanding communications networks within a firm allows a more comprehensive understanding of its governance and hence, its culture.

Organisational culture is about the beliefs and values that people hold about a company (see Deal and Kennedy, 1982; Schein, 1985). It is intangible and in many respects implicit within the behaviour of an organisation's employees. Thus, the patterns of communication found within organisations are not rendered easier to understand by virtue of the boundaries that define them. The construction industry is in many ways an exemplar of this problem, as it has a stubborn history of communication problems between its culturally diverse and organisationally fragmented occupational groups that seemingly exist even within firms (Crichton, 1966; NEDO, 1988; Latham, 1994; Moore and Dainty, 1999, 2001). It is important, therefore, that those seeking to understand an organisation are able to deconstruct the processes which together constitute its operation and in particular, that define its communication patterns and flows.

A systems view of organisational communication

Systems theory is a powerful tool in understanding organisational communication because it highlights the importance of communication in organising (Eisenberg and Goodhall, 1993: 105). The 'systems' view of organisations is a powerful metaphor as it helps to address the interrelationships of structure *and* behaviour within them (Mullins, 1999). The open systems view sees an organisation as a combination of interdependent parts (sub-systems) which collectively make up the whole (or in metaphorical terms as a living organism which has to adapt continually and naturally to changes in its environment in order to survive). The value of systems theory to the study of organisations is its ability to simplify complex situations by considering its sub-systems and the relationships and interdependencies between them. An open-systems view suggests that an organisation takes inputs/resources from its environment, transforms them in some way and then sends outputs back to that environment. In construction, it is a perspective which has proven useful in explaining organisational effectiveness and for understanding why organisations succeed or fail (Loosemore *et al.*, 2003). A criticism of the systems view, however, is that it is the people, rather than the organisations themselves, that react to the environment (Silverman, 1970: 37). It is important to recognise, therefore, that attempting to view organisational life in a 'rational' manner is likely to oversimplify the complexities of their functioning. Thus, although adopting a systems view of organisations allows an improved understanding of how communication flows within them, it does not necessarily inform us of the *nature* of the dialogue itself (Eisenberg and Goodhall, 1993). Nevertheless, it remains a good starting point for understanding communication within construction firms.

Formal and informal aspects of organisational communication

As was discussed earlier in this chapter, communication can be formally or informally defined within organisational settings. Informal communication practices within organisations have been widely examined within the mainstream management literature. Krackhardt and Hanson (1993) provide an excellent metaphor for distinguishing between the formal and informal system, the formal being the 'skeleton' of a company, and the informal being the 'central nervous system' which drives the collective thought processes, actions and reactions of an organisation's business units. This analogy is useful as it can be recognised that both are necessary for the effective functioning of the firm; removing either aspect renders it useless. It is perhaps the evolution and adaptive nature of these processes which lend credence to the concept of the 'evolving structural' perspective. Fisher (1980) suggested that communication should not be thought of as a structural entity, but more as a sequence of events which occur over time. The sequences become familiar as certain reactions tend to follow specific acts, and are repeated so often that an organisation's actors come to expect the next act in sequence even before it occurs. To further emphasise the point, Wofford *et al.* (1977) use the analogy of a learning curve of a football team, in which early season performance tends to be exceeded later in the season, as team members' ability to 'read' one another improves. This view is arguably supported the proponents of strategic partnering who see benefits accruing in relation to the performance of a construction team as a 'synergistic' relationship develops between the various participants.

In his research into communications, Hill (1995) also makes a useful observation regarding the relative effectiveness of formal and informal communications. A participant in his study mentioned that informal communication was what 'got the job done'. It was noted that the description was effectively borrowed from a formal understanding of the purpose of an organisation. Ironically it appeared that the informal operation of communication fulfilled the explicit objectives of the formal system – which in turn indicated that the formal system set up was actually incapable of 'delivering the goods' for the organisation. Work by Dulaimi and Dalziel (1994) also found interesting behaviour in a comparison between the management synergy in design and build projects and those procured under traditional means. They found that communication was in general more informal, frequent and satisfactory in design and build projects. This research can be used to validate the results of earlier work conducted by Pinto and Pinto (1991) who found that high cooperation teams tend to use more informal communication channels (telephone and informal discussion) as compared to low cooperation project teams. Such studies underline the importance of effective informal communication channels in the functioning of high performing projects and organisations.

One mode of informal communication understood by all organisational participants is known as the 'grapevine'. It is probably familiar to most people who have been employed for a large organisation where rumours and information flow between members in an informal manner and without regard to management hierarchy. Although grapevine communications are generally considered to be informal and unregulated, Davis (1953) and Schwartz and Jacobson (1977) have found that these communications can be represented by 'grapevine' charts within which particular patterns of communication emerge. The term tends to have a negative connotation and this may be a result of its association with rumour and gossip within organisations. However as Davis and Newstrom (1985) point out, the original use of the term may also have contributed to this view. It derives from the use of telegraph lines (used during the American Civil War) that were strung from tree to tree with wild grapevines growing over the lines in some areas. The messages that were carried along these communication channels were often distorted and garbled. However, a contrary view is offered by Wofford et al. (1997) who argue that the grapevine is an example of how people use informal communication systems to assist in the achievement of organisational work. Certainly, it is the normal mechanism used in story-telling; an activity that can provide distinct organisational learning benefits.

Communication networks within construction organisations

When the formal and informal aspects of communication are understood, an appreciation can be gained of the 'networks' which exist within the organisation. An effective way to conceptualise organisations is as collectives of individuals who work in groups (Hawes, 1974). A group can be defined as two or more people who consider themselves to be connected in some way and who must therefore communicate over time. As has been discussed earlier, these groups can be formal in nature (such as colleagues working within the same department) or informal (such as like-minded individuals from different organisations coming together to share good practice for their own mutual benefit). The patterns of communication which emerge between these groups are to some extent determined by the communication 'network' (the groups that coexist within the boundaries of the firm). There are many factors that affect the ways in which these groups interact, not least the way in which managers design them, their geographical location and their lines of authority (Eisenberg and Goodhall, 1993: 271). However, regardless of how they are configured, they will inevitably play a vital role in determining how information flows between people and groups. Understanding communication networks is therefore vital to understanding the ways in which organisations function.

As was discussed earlier in this chapter, formal networks are those defined by the organisational structure, rules/procedures and recognised relationships between people, teams and functional departments. Although these are important, the most powerful and important groups in organisations are not appointed at all, but emerge from the communication that people undertake as part of their organisational life (Eisenberg and Goodhall, 1993: 273). These emergent networks coexist and interact with the formal networks. As was discussed earlier, there is evidence to suggest that communication has a more powerful bearing on structure than has structure on communication (Rogers and Agarwala-Rogers, 1976). For example, some departments or groups may have no formal relationship and yet may need to communicate continually in order to control their workflow. Consider, for example, two construction projects situated close together but managed by different divisions of the business. Although they may operate as part of completely disparate profit centres within the organisation, they may wish to maintain a dialogue in order to share resources or to benefit from the economies of scale that collaboration could generate. Thus, an informal and temporary alliance may ensue through which both parties benefit, but which lies completely outside of the formal structures defined by the organisation.

In construction firms, organisational networks vary in size and density. In this context 'density' refers to the total number of connections with other groups. Dense groupings (or 'clusters') of individuals have a profound impact on innovation within organisations and determining whether others will adopt new processes or technologies (see Albrecht and Hall, 1991). This is one reason as to why communities of practice are known to be organisationally beneficial to the generation and adoption of new ideas (see Wenger et al., 2002). It is incumbent upon managers to encourage and foster dense clusters of individuals to come together if they wish to benefit from the improvements that this can generate. Managers should identify the more successful informal networks in order that they can encourage them to grow and develop for the benefit of the business. These processes are embodied within the process of defining the organisation's structure and culture.

The role of communication in managing organisational change

Understanding organisational communication demands that the project manager also appreciates its fundamental importance in allowing it to cope with change. For construction companies, change is an inevitable consequence of organisational life. This is because organisations must respond to continually evolving market threats and opportunities (manifested in customer demands), regulation, resources and other events which combine to shape strategy and action (Daly *et al.*, 2003). Within business generally, it is estimated that around 70 per cent of change programmes end in failure (Senge *et al.*, 1999), primarily because of poor internal communication (Murdoch, 1997). This suggests that construction organisations should consider the way in which change is communicated before any programme of change is implemented. Failure to do this will lead to a rejection amongst the employees expected to embrace and work within new situations and structures. In contrast, effective communication can enable the seamless integration of even radical change initiatives within firms. Schweiger and De Nisi (1991) found that when managers are open and honest to employees, even when conveying bad news, this resulted in less absenteeism, higher levels of productivity and lower staff turnover. Thus, open communication is an important enabler of organisational change management (Quirke, 1995).

From a project perspective, Sidwell (1990) acknowledged that construction projects have been compared to living organisms, in that they change in structure and style at each stage of a project lifecycle. He suggests that project organisations continually transform through organic, mechanistic and bureaucratic structures. Indeed, Sidwell concludes that the ability to metamorphose within the project environment contributes towards a successful project outcome. It may be the case, therefore, that change is easier to manage in project-based organisations such as construction companies because they are more used to coping with the change defined by their temporal nature.

The nature of communication within construction organisations

In order to understand the factors which impact upon the effectiveness of communication within construction companies, it is first necessary to explore the aspects of organisations which define them. As was alluded to above, in simple terms, the effectiveness of communication within an organisational setting is contingent upon three sets of interrelated and mutually dependent variables:

- the structures protocols and procedures which define the ways in which information is channelled;
- the norms which shape the nature of communication and information flow within the organisation (or the organisational culture); and
- the attitudes, behaviours and actions of the people responsible for communicating and acting upon the information transferred.

This perspective recognises the existence not only of both formal communication channels and informal networks, but also the power of the individual in determining how information is acted upon and translated in practice. Each dimension is explained in more detail below.

Communication structures within construction organisations

Given the vast information needs of construction projects and their participants, robust communication channels are vital for the effective functioning of the industry's organisations. These include protocols on roles and responsibilities in terms of relaying relevant information, as well as the form in which it is to be moved. Different functional departments, groups and individuals will have their own information needs that will, in turn, generate their own information outputs. By defining a communication structure (i.e. a formal representation of how different groups and networks are interconnected through their information needs and flows) and supporting this with communication protocols (e.g. standard proformas for transferring information from one department to another or electronic data exchange standards to ensure the compatibility of information moved between different computer systems), more efficient use and more rapid understanding and utilisation of the information transferred can result.

At a project level, many construction organisations define formal information management roles within the project team. This can include roles for controlling the collation and subsequent distribution of documents such as drawings and specifications, the modelling of data on computer-aided design or visualisation packages and the utilisation of cost and programme information with implications to the production effort. Although it is crucial for ensuring that information is efficiently managed, a wholly structural view is very limited as it basically views the organisation as a series of channels for the transfer of information (and hence sees communication as merely an 'information transfer' mechanism). This oversimplifies the organisational communication process as it views the receiver as a passive receptor, uninvolved in shaping the meaning of the message. Thus, on its own, a structural perspective cannot account for why organisational communication succeeds or fails. This is particularly important in construction where the production effort involves many people from different functional, cultural and organisational backgrounds who are likely to respond in different ways to the same communication media.

Kerzner (1997) describes how organisational charts can also be used to describe how internal and external communication 'should' take place and recommends that project managers employ a communication responsibility matrix. However, as previously stated, the UK construction industry

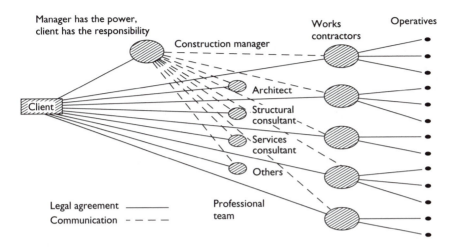

Figure 6.1 Communication responsibilities in the construction management.
Source: Moxley (1993).

appears to rely on less formal mechanisms of project management. Kerzner acknowledges that organisational charts are a valuable tool for management, but do not necessarily describe how people interact within a project programme. He cites Karger and Murdick (1963) to suggest that just because the chart represents something as fact does not necessarily mean that it is so in reality. Academic writers such as Moxley (1993) have attempted to conceptualise the communication network within projects. Moxley provides a communication network for the construction management procurement system (see Figure 6.1). It assumes a primary assessment of the project from the client and construction manager perspectives and no communication links are shown between the design team, design team to works contractors and one work contractor to another. However, it does convey how the formal (legal or contractual) linkages generated by a project organisational system bear little resemblance to the informal (communication) network which surrounds it. It can be appreciated that the realities of the communication patterns that emerge from such an organisation are exceptionally complex.

Culture and communication within construction organisations

The culture metaphor for organisations is a reaction against the mechanics of complex systems, as it shifts the focus to the language of the workplace, the routines of employees and the shared practices that make an organisation unique (see Eisenberg and Goodhall, 1993: 115). Culture shapes everything that people do in organisations, especially the way in which they communicate with each other. The operation of most construction companies is grounded in a strong culture brought about by powerful personalities at the top of the organisation. This can apply as much to a large business as it can to a small firm.

Deal and Kennedy (1982) suggest that there are five key elements to a strong culture: (i) *the business environment* in which the organisation operates; (ii) the *values* of the organisation and the commitment to them; (iii) *'heroes'* who have benefited the organisation in some way and so act as role models for others; (iv) *rites and rituals* which define the ways in which members feel about themselves and the organisation; and (v) *the cultural network* of formal and informal communication channels. These five dimensions come together to form the cultural context of the firm, which will in turn determine *how* communication takes place and how effective it is. For example, consider a small construction firm led by a charismatic and entrepreneurial owner/manager with a rigid but historically effective approach to managing his business. This individual is likely to instil a set of values and ideals within his staff which he sees as congruent with meeting the demands of the business environment in which they operate. The cultural network which evolves from such a situation is likely to reflect the dominant ethos of the owner/manager and hence, reproduce the cultural norms of the leader. It is for this reason that critical perspectives on organisational communication tend to focus upon the exercise and abuse of power within them (Eisenberg and Goodhall, 1993: 159). Cultural influence on communication can be so powerful that it can override the structural aspects designed to ensure the effective use of information within it.

Individual behaviour within construction organisations

The key to successful communication is to ensure that what is communicated is purposeful. This means that there should be communications for sharing in the vision or mission of the change, integrating the efforts of those involved *and* securing their commitment. Although structural mechanisms and cultural values and norms are important to the impact that communication has within an organisational setting, it must nevertheless be meaningful to the individual if it is to be considered truly effective. As was examined in Chapters 3 and 4, the way in which the individual interprets information is highly personal and context specific. Hence, no matter how well structured a business is and how well aligned with the pervading culture, individuals can still ignore or misinterpret information that they receive. This perspective recognises that discourse within organisations is essentially a social activity where members use and interpret language to structure their own realities (Francis, 2002).

Individuals may not communicate with others effectively because of factors grounded in their own personality or prejudices which lie outside of the cultural environment of the organisation and which cannot be mitigated by structural communication channels. In addition, the communication role played by individuals may differ fundamentally from their formal job role within the organisation. This is because, for most people, their individual communication network is something that they value and hence, nurture, develop and maintain to their own advantage. Individuals from particular professional backgrounds will have their own role-defined language and behaviours which are as likely to affect their behaviour as their sense of organisational identity. Moreover, the impact of such behaviours and attitudes are not necessarily determined by seniority, as even employees of relatively junior hierarchical standing can find themselves in very powerful communication positions. For example, the secretary of the Managing Director of a major construction company could be party to more strategically significant information than even senior managers within the firm.

Communication and decision-making in organisations: a contingency view

Group communication is more frequently directed toward group decision-making than toward any other type of objective (Wofford *et al.*, 1977). However, much of the research into how group communications function to affect group decision-making has lacked a viable frame of reference and has failed to provide clear evidence on the role that communication plays in successful and unsuccessful decision-making. Gouran and Hirokawa (1983) suggested that communication functions may affect the quality of group decisions positively or negatively, and can even disrupt or counteract decisions if not managed appropriately. It would seem that there are no magic formulas that a group can employ to ensure good decisions (Fisher, 1980). Indeed, if we consider the concept of 'groupthink' as espoused by Janis (1972) (see Chapter 5), it can be appreciated that group decision-making can have negative impacts on human performance. It would seem, therefore, that different types of organisational communication will be required depending upon the demands of the situation. Their effectiveness will be contingent upon the existence of flexible decision-making frameworks and the robustness of group decision processes.

Several studies in the organisation of construction projects literature advocate what could be described as a rational approach to the design and management of an organisation (Walker, 1980; Hughes, 1989; Cornick and Mather, 1999). However, such authors also recognise the importance of construction project managers' experience and intuition in selecting appropriate solutions. Laufer *et al.* (1999) argues that project managers should manage the team's decision-making, and not make the decisions themselves.

This suggests that 'successful' project managers will employ some form of decision matrix which formalises the decision-making process so that project participants will know when and how to input. Laufer *et al.* (ibid.) conclude that the mere listing of decision items and decision-making roles can quickly help to create order and certainty within project teams. This view clearly sees the project manager's role as that of a decision-making facilitator, but it could equally be argued that project managers who are the culprits in unsuccessful decision-making events should not fulfil such a role. This discussion emphasises that there is no best way to manage decision-making in organisations, particularly in dynamic project-based environments like those formed in construction. Rather, the focus should be on allowing the members to develop appropriate decision frameworks which respond to the prevailing situation and for those in the best position to decide on the best course of action to do so.

An examination of the literature surrounding the optimal conditions for effective decision-making to prosper reveals some fundamental contradictions. For example, Holloman and Hendrick (1972) found that decision-making procedures which require and/or permit increased social interaction produced better decisions than did procedures involving less interaction. In contrast, Hirokawa (1980) found that group members in effective decision-making groups tended to produce more procedural statements than members of ineffective groups. In addition they also tended to consistently spend more time engaging on procedural matters. The findings of these studies emphasise the fact that different groups will require differing levels of formality in order to manage their decision processes effectively. New groups may require more by way of protocol and procedure than those familiar with working together. Thus, the level of formality and external control required will be to some extent determined by the knowledge of the work group of their fellow members, as well as by the urgency of the decision-making process and the importance of the decisions in hand.

Managing communication within construction organisations

For most project-based managers reading this text, their focus of interest is likely to be on how to communicate effectively with their employees in a way which encourages them to contribute positively to the production effort. However, despite the problematic communication context of the industry's intra-team and inter-team communication environment, very little guidance is available that describes how to improve such processes within the construction sector (Cheng *et al.*, 2001). This is surprising given that the degree to which those engaged in productive activity are motivated depends upon the effectiveness of their managers to communicate with them (Armstrong, 2001: 807). In response to this need, the basic building blocks of an effective communication strategy applicable for the temporary organisation of the construction project are outlined below. This should not be seen as a prescriptive or normative checklist for enabling successful project outcomes, but as a broad framework of principles that can be used to help shape a communication strategy contingent with the specific needs of the project in hand and with the nature of those involved with the endeavour.

Understanding the prevailing structural environment

A fundamental first step in defining an organisational communication strategy should be to understand the prevailing structure and underpinning communication network which arises from it. Without such an understanding it will be impossible for the project manager to decide upon how and where to introduce changes in their organisational approach to communication (Weinshall, 1979). In attempting to chart the organisational system, several mapping techniques are available. These maps are known as an 'organigram', which can be compared to a map or an aerial photograph depicting linkages between individuals. The Formaligram and Informaligram (also known as a Sociogram) describe the relationships as viewed by the organisations participants. These are useful insofar that they can highlight disagreements and omissions between people as to their roles relative to each other. Weinshall (1979) also developed a conceptual tool – a communication chart or communicogram – used to compare aspects of interaction among management employees. Reference to structural terms such as 'organigram' and 'formaligram' appears lacking in the construction management field, despite a clear need to 'organise' a project or a business. However, Hellard (1995) acknowledges that 'conscious attention' must be paid to organisational relationships within a construction team. He recommends that organigrams should be produced and that responsibilities, contractual relationships and communication channels should be made explicit to all project participants (See Case Study in Chapter 2). This recommendation for organisational transparency has also been emphasised by other construction management researchers (e.g. Walker, 1980; Hughes, 1989).

Recognising the role of the HRM function

The increasing acknowledgement of the importance of communication in organisations is reflected in the rising prominence of communications in the HRM literature. Whereas the personnel management paradigm emphasises a restricted flow of indirect communication, the Human Resource Management new orthodoxy emphasises an increased flow of *individual* communication with the workforce (Storey, 1993). Thus, whereas in the past the focus has been on conveying information to employees (down) through an organisational hierarchy, the focus in the HRM literature in recent years has been on *upwards* communication from employees to their employers. According to Torrington and Hall (1998: 114), upwards communication is vital for:

1 Understanding employees' concerns.
2 Keeping in touch with employees' attitudes and values.
3 Alerting managers to potential problems.
4 Providing managers with workable solutions to problems.
5 Providing managers with information necessary for effective decision-making.
6 Encouraging employees to contribute and participate with organisational decision-making, thereby improving motivation and commitment to organisational actions and directions.
7 Providing feedback on the effectiveness of downwards communications.

Although construction has long been criticised for its antiquated approach to many aspects of the HRM function (Langford *et al.*, 1995; Loosemore *et al.*, 2003), elements of this approach can be seen to have permeated HRM in the industry. For example, whereas in the past information flow has been mediated by employee representation or trade unions, contemporary approaches have placed an emphasis on communicating directly with the workforce via individualised employment contracts. Indeed, the decline in the significance of collective bargaining is likely to lead to an increased emphasis on direct communication in the future (Emmott and Hutchinson, 1998). Thus, strategies for communicating within construction organisations tend to reflect the new employment paradigm and the shifts in workplace culture that this has engendered.

From mid-2005, there will be a legislative imperative (embodied in Directive 2002/14/EC of the European Parliament) for employers to consult with their workforce collectively. By 2008, the directive will cover all employers (private and public) within the United Kingdom employing 50 or more persons. The directive gives employees a right to be informed about the business's economic situation, informed and consulted about employment prospects, and informed and consulted about decisions likely to lead to substantial changes in work organisation or contractual relations including redundancies and transfers. Beaumont and Hunter (2003) note that the general thrust of the directive is for employees to engage in dialogue with their employees via collectives rather than by individual means. Information and consultation has to take place at an appropriate time and at the relevant level of management. Normally it will be done via an employee representative, defined according to national law and practice. It would seem, therefore, that it is incumbent upon organisations to define broad channels of communication to comply with the consultation legislation whilst simultaneously supporting managers in maintaining an individual communications dialogue with their staff teams.

Measuring communication effectiveness in construction

Being able to benchmark existing communication effectiveness is a prerequisite to identifying where improvements in performance can be made. It is essential that all managers within the industry continually appraise the success of their communications policies and approaches in order to establish whether they are suited to the environments posed by new projects and project participants. In establishing methodologies for the industry to achieve this, a useful starting point is provided by the outcomes of research undertaken by the Construction Industry Institute (CII) in the United States (see Tucker *et al.*, 1997). This has led to the development of a project management 'toolkit' for measuring project communication effectiveness. The tool was developed as part of the CII research into the potential for enhanced project success through improvements in project team communications. Development of the tool was in the context of use by project managers as a diagnostic tool to assess project team communication effectiveness during the design and construction phases of an engineering, procurement and construction (EPC) project. Software is used to consolidate and analyse survey data, which then reports results in an overall communications effectiveness score. The CII developed the toolkit after reviewing communication on 72 projects. Six categories of communications were identified as having a direct impact on participant's perceptions of project success. Statistical analysis of the CII data revealed critical communication effectiveness variables (see Table 6.1). The weighting factor applied to each category was developed as a means to reflect the categories relative importance for effective communication. Thomas *et al.* (1998)

Table 6.1 Critical categories of communication

Category	Description	Weight
Accuracy	The accuracy of information received as indicated by the frequency of conflicting instructions, poor communications, and lack of coordination	2.1
Procedures	The existence, use and effectiveness of formally defined procedures outlining scope, methods etc.	1.9
Barriers	Presence of barriers (interpersonal, accessibility, logistics, etc.) impeding communications between supervisor or other groups	1.8
Understanding	Understanding information expectations with supervisors and other groups	1.6
Timeliness	Timeliness of information received including design and schedule changes	1.4
Completeness	The amount of relevant information received	1.2

Source: Thomas *et al.* (1998). Reproduced by kind permission of the *American Society of Civil Engineers.*

argued that their study represents a milestone for engineering and construction projects in that it has identified and measured critical performance variables.

The CII approach offers a convenient methodology for establishing any failings in the current level of organisational communication effectiveness, especially if used in conjunction with measures of outturn performance likely to stem from an effective communication climate.

Developing an organisational communication strategy

Organisations require different approaches to their employee communication which are contingent upon their size, culture, management style, resources, staff and market conditions (Kitchen, 1997). For example, the communication needs of a very small (or micro) sized construction firm focusing on the local or regional market will be markedly different from a large organisation operating globally. Thus, formulating a communications strategy first demands a thorough analysis of what management wishes to communicate, what employees want to hear and the barriers to conveying and/or receiving information effectively (Armstrong, 2001: 809). These analyses should reveal the nature of the organisational communication requirements that will, in turn, allow appropriate communication systems (or channels) and HRM policies to be designed.

Organisation communication media

Any good organisational communication strategy should comprise a series of interrelated communication systems, themselves comprising a range of different media designed to ensure that information flows effectively and that the message is understood as was originally intended. Examples of communications media at the disposal of managers and commonly used in organisations include:

- *Intranet systems* – most medium and large organisations have intranet systems and/or access to electronic mail which allowed information to be delivered rapidly to a wide audience. Considering that this is a cheap and efficient way of transferring information, it is particularly suitable for organisations that have to rapidly communicate a large amount of information. It also allows employees to feedback to managers using the same systems. The downside of these types of approaches is that employees need access to computers to access the information and information technology skills to be able to retrieve it. Other issues in relation to IT-based communication systems are discussed in Chapter 8.

- *Magazines and newsletters* – these formal mechanisms are used by many organisations for both public relations and internal information purposes. Newsletters tend to be more internally focused for employees and may include information on the performance of the organisation and opportunities arising within it.
- *Notice boards* – although a fairly crude way of conveying information, most organisations make extensive use of notice boards to publicise general information. Obviously there is no guarantee that employees will read what is posted and information quickly goes out of date, but it remains a cheap and relatively efficient communication mechanism.
- *Team briefing* – this involves communicating to small teams of employees and giving them the opportunity to receive, discuss and feedback on issues of relevance to the organisation. By developing a chain of communication through the various hierarchical levels of the organisation, the idea is that information can cascade down through the organisation and queries be answered on issues pertinent to those operating at particular levels within the hierarchy. Such systems may be particularly appropriate for project-based organisations for overcoming the inherent difficulties created by the distributed nature of construction sites.
- *Focus groups* – these are increasingly being used by organisations to bring together people from various levels of an organisation to discuss issues of cross-cutting relevance. By promoting open debate around the issue at stake and inviting a cross-section of opinion, a more rounded and complete view of the impact of new policies can be made, and the buy-in of those at all levels within the organisation secured. Care must be taken to involve as many people as is practically possible in order to ensure that employees do not feel disenfranchised by the process.
- *Staff surveys* – these are usually carried out by questionnaires (sometimes anonymously) in order to collect the views of employees on issues affecting them. These can be an effective way of ascertaining the impact of organisational policy on employees, but depend upon an adequate cross-section of them responding for the results to be meaningful.

It is likely that a combination of these strategies will result in an effective communication system. It is important to ensure, however, that where more than one medium is utilised that the combination conveys messages in a mutually supportive way. If different mechanisms convey different meaning behind the messages sent then this may result in less effective communication than relying on a single channel.

HRM policies to support effective communication

Communication forms an enabling element of many aspects of the HRM function within contemporary organisations. For example, communication is an important factor in motivating the workforce, since it is essential to keeping people feeling informed, happy, valued and involved. It is also essential for preventing misunderstandings which are often the root cause of industrial relations problems. This is particularly important within a construction project context, as communication has an important role to play in influencing general employee opinion on the levels of openness and involvement within the wider organisation; an open communications culture is more likely to foster an egalitarian and trusting culture at all levels of the business. Finally, good practice in many areas such as occupational health and safety, equal opportunities and training and development are also reliant on effective communications between key stakeholders (Loosemore *et al.*, 2003). Two important HRM policy areas that can be used by construction organisations to foster open and effective communications with their employees are reward-structure approaches to participation and empowerment:

- *Performance Management and Reward* – reward systems communicate important messages to employees about what the organisation values from its employees and so are powerful tools in encouraging employees to communicate in a way which supports organisational objectives. Managers can be helped in their communication responsibilities by systems for reward and/or sanction which encourage the types of behaviour considered desirable (Huczynski and Buchanan, 2001: 200). For example, if site-based employees are rewarded for sharing knowledge and information on new and innovative ideas and techniques (as part of a formal knowledge management policy), then employees are likely to place more importance on this activity. Similarly, if employees are encouraged to communicate across their professional groups for the good of the project and are rewarded for the improved performance that results, a culture of effective communication within project teams may result which has continued benefits in the future.

- *Employee Involvement, Participation and Empowerment* – the concept of employee involvement, participation and ultimately empowerment has grown throughout the 1980s and 1990s, largely in response to reduced union power and government policy which encouraged firms to evolve arrangements which suited their needs (Wilkinson, 2001). Enabling employees to become more involved in controlling the work that they undertake is an effective way of optimising their contribution to the organisation, especially in terms of communicating effectively. Management initiatives used to engender an 'involvement climate' include quality circles, customer service improvement schemes and business process reengineering (BPR). These have become popular in the construction industry in recent years because they help to ensure that employees' opinions and perspectives influence the direction and policies of the organisation.

Some caveats to improving communication within the temporary involvement climate of the construction project

This chapter has defined a clear need for a communication system which combines both formal (structural) and informal (cultural and interpersonal) communication mechanisms if employees are to feel informed, motivated, involved and ultimately empowered in a way that enables the organisation to cope with the inevitable change that it will confront. However, it is important to understand that establishing effective communications is not just about how information is transmitted from one person to another, but also about the ways in which it is interpreted, understood and then acted upon (see Chapters 3 and 4). This presents a particular problem within the construction project environment where the distributed and temporal nature of project teams undermine direct control and influence over the way in which information is managed and interpreted. Indeed, the dichotomy between temporary and permanent organisation structures and the use of formal organisational charts was noted by Halsey and Margerison (1978), who suggested that theories in mainstream management text are inappropriate for the design, management and conceptualisation of construction project organisation.

Direct and open communications have not traditionally formed a characteristic of construction projects, which as a result of conflicting interests and penal contracts, are traditionally characterised by secrecy, top-down information flows and manipulative behaviour. A hallmark of the matrix organisation (which characterises the operation of the majority of construction firms) is the prevalence of project teams where members have both functional and project responsibilities (and hence, lines of authority) (Scott, 1981). No matter how effective the project manager is in managing communication within the project environment, employee dissatisfaction will result if this is not supported with effective communication from elsewhere within the organisation. Cross-functional teams therefore pose their own communication issues and challenges within the construction organisation.

An additional layer of complexity is provided by the fact that construction project communication is an inherently multi-organisational phenomenon. This is because many different parties will be involved in the collective effort required for any project. This has led to a fragmented delivery system which now usually comprises of many parties with an extended and unwieldy supply chain. The coordination of the inputs of these parties and management of production information inherently relies upon both formal and informal communication practices (Emmitt and Gorse, 2003: 91). As such, planning communication in such a way that combines formal and informal exchange of information is vital for achieving project objectives. Formal aspects of communication are often determined by the legal conditions of contract formally agreed and set out before the project commences.

However, if only these were relied upon, then construction activity would be so litigious that normal working relations would probably never develop! This has long been a failing of the construction industry as was recognised by both the Latham (1994) and Egan (1998) reports which advocated greater trust, partnerships and alliances within the supply chain. Engendering this within the temporary organisation of the project team, however, remains extremely problematic.

It could be reasonably expected that the temporal nature of construction teams would lead to discontinuities as members attempt to overcome professional and functional boundaries within them. However, on some projects, the pressures, cohesion, loyalties, focus, and momentum that can develop become so intense that they effectively seal themselves off from the outside world, considering outsiders as an unnecessary distraction and even covering up problems that may expose internal weaknesses to them (Loosemore, 2000). Thus, communication difficulties can also occur between the project team and their external environment, even when it is made up of individuals from different organisations. All of these possibilities must be considered by managers when designing communication structures to connect their disparate project teams.

Summary

This chapter has emphasised the highly complex and multi-dimensional nature of organisational communication. This renders its understanding extremely problematic, particularly within the temporary organisation of the construction project. Fortunately, the study of intra-organisational communication is probably supported by the greatest amount of theory and empirical insight. The systems perspective in particular facilitates understanding of the importance of construction organisations adapting their communication processes in a way that recognises specific pressures and the nature of the individuals that they employ. This views the organisation as comprising both internal and external influences, both of which will impact on communication within it. Internal influences include the structure and culture of the company, but both can be moderated by the individual perspectives of the actors within them. Adopting an open systems or contingency view also recognises that there is no single effective way to manage communication within construction organisations. Indeed, when formulating an organisational communication strategy it is essential that the individual circumstances of the organisation are taken into account. Specifically, it is important to recognise the impacts of formal structure, informal culture and the individual actions of the organisation's members when attempting to understand communication within them. In terms of aligning the needs of employees with those of the organisation, it is also incumbent upon the organisation to ensure that communications systems are aligned with HRM policies and vice versa. This is particularly important within the distributed environment of construction projects where policies in reward management and involvement help ensure a consistent and understood set of policies which promote engagement and conformity to organisational goals. The studies and perspectives presented within this chapter emphasise that project managers must remain acutely aware that communicating within such a problematic organisational environment demands that they continually change and modify their approach to enable them to respond to the dynamic nature of the construction project. They have also emphasised that it is only through the combination of a formal communications structure supported by effective informal communications practices that sufficiently flexible and robust communications are likely to result.

Critical discussion questions

1 A great deal of knowledge can be gleaned about an organisation's culture by identifying the discrepancies between its informal communications network and its formal organisational structure. Compare and contrast the formal and informal communications patterns and 'grapevines' that exist within an organisation with which you are familiar. Evaluate the ways in which the informal communications of the organisation support and/or impede the formal communications hierarchy and structure (as depicted by the organisational organagram).

2 An architect has a particular vision for the design and specification of a building which is grounded in his/her personal interpretation of the client's brief. However, this design does not accord with the favoured design approach of the practice for which they work or the other parties who must translate the scheme into a practicable design. Devise a communications strategy for ensuring that the perspectives and interpretations of this individual do not undermine the overall success of the project from the organisation's perspective.

Case study: community relations at HBG Construction Ltd: internal and external communications

Introduction

This chapter has revealed the highly complex and interdependent nature of formal and informal communication within an organisational context. Managing organisational communication effectively is difficult for construction firms as they must continually ensure that their employees communicate effectively with each other *and* with the external parties with

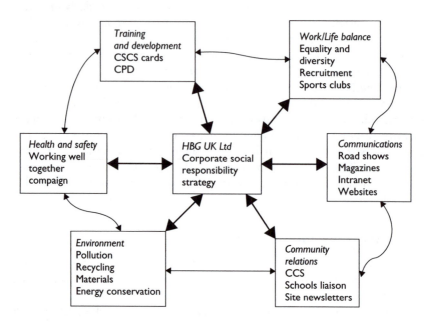

Figure 6.2 CSR strategy at HBG UK Ltd.

whom they interface. This case study presents an excellent example of how internal and external communications can be managed effectively by a construction firm. It focuses on two aspects of the organisation's Corporate Social Responsibility (CSR) strategy – Community Relations and Communications. These two aspects, combined with the five other parts of their CSR strategy (see Figure 6.2) provides the company with a 'brand value' that is attractive to potential employees, customers and project partners. The seven individual areas are managed through individual policy documents that are themselves determined by and, indeed, influence, the overall CSR strategy. The methods used to communicate the CSR vision from the boardroom to site level so as to ensure the adoption of community relations practices in all of their projects represents a major challenge for a large construction company. It demands effective internal and external communications practices, as the case study demonstrates.

HBG's corporate social responsibility programme

HBG UK Ltd is part of the Royal BAM Group (Koninklijke BAM Groep nv) which has its headquarters in the Netherlands. Royal BAM Group ranks among the largest construction firms in Europe with around 27,000 employees and a turnover of approximately 7 billion (Euro) in 2003/2004. In the United Kingdom, HBG's £800 million turnover is generated from an integrated range of services including construction, property development, design, private finance initiative (PFI) and facilities management work.

In November 2002 the HBG CSR Policy was launched throughout the company. Copies of the policy, accompanied by a letter of explanation, were delivered to the home of every member of staff. In addition, posters were put up in every office and site throughout the country. This preceded the policy distribution and continued after its launch. The six elements that underpinned the programme were: environment and sustainability, training and development, health, safety and welfare, community relations, communications, and work/life balance. In addition, the company also developed 'model' project-specific CSR plans for several projects to serve as templates for future project planning and submissions. Figure 6.3 shows the

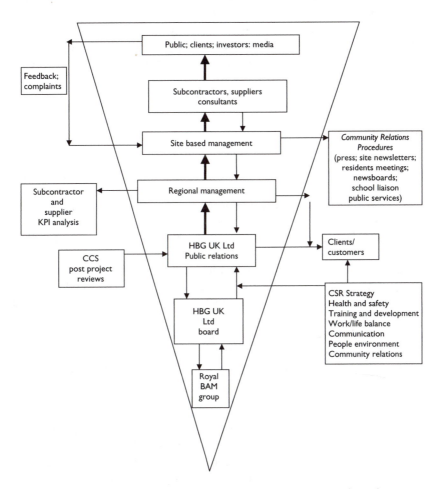

Figure 6.3 CSR at HBG UK Ltd: internal and external communication channels.

communication channels for the dissemination and implementation of the CSR policy throughout the organisation in the United Kingdom. The Public Relations (PR) Department has a key role in participating at boardroom level and influencing senior members as to the importance of CSR. In addition, the PR team have a fundamental role in coordinating the seven strands of the CSR strategy shown in Figure 6.3. However, it is clear that site-based management also have a crucial role in communicating the CSR policy to local communities and for applying the eight-point CCS code of practice.

The considerate constructors scheme

An integral aspect of HBG's CSR policy was to adopt the code of practice set out within the CCS. The CCS is a Construction Industry Council (CIC) run initiative which aims to promote a positive image for the construction industry image at a site level (see Barthorpe, 1999 for more details on this scheme). In May 2004 HBG signed up to become a signatory of the code and in May 2004 became an Associate Member of the scheme throughout the United Kingdom. Table 6.2 shows details of the CCS code.

CCS at HBG

Mr Alan Smith, HBG's Director of Public Relations, argued for the whole-sale adoption of the CCS at HBG. In May 2003 the Board at HBG Construction Ltd issued instructions to all HBG Regional and Business Unit Directors for all of their construction sites to be registered under the CCS or another equivalent scheme. The company recognised that this policy would provide them with an unrivalled 'shop window' to display their brand and operations nationwide. However, they also recognised that joining such a scheme would also make them far more vulnerable to criticism. As such, after the first full year of operating the CCS, HBG conducted an intensive review of their own performance under the eight categories in the CCS scheme code. The results of this survey were published in hard copy (HBG, 2004a) and distributed to all regional and site-based staff, as well as being posted on the company intranet.

The analysis of HBG's performance under the CCS eight-point code was conducted by the company's Public Relations Department. In addition, the company has eight regional coordinators, each of whom acted as champion for the company's commitment to the scheme. Results showed HBG well above the industry average in every region in the United Kingdom. As a result of the review, it was found that not all of their sites were signed up to the CCS scheme and new procedures were established so that registration forms are completed immediately on contract award with copies sent to both the regional and corporate scheme coordinators. The need for

Table 6.2 CCS eight-point code of practice

Consideration
All work is to be carried out
  with positive consideration to the
  needs of traders and businesses, site
  personnel and visitors,
  pedestrians, shoppers and general public.
  Special attention is to be given to
  the needs of those with sight, hearing
  and mobility difficulties

Environment
Noise from construction operations
  and all other sources is to be kept to
  a minimum at all times. Consideration
  should be given to the selection
  and use of resources, utilising local
  wherever possible. Attention should be
  paid to waste management and the
  avoidance of pollution – recycling of
  surplus materials is encouraged

Neighbourliness
General information regarding the
  scheme shall be provided for all neighbours
  affected by the work. Full and regular
  communications with neighbours, including
  adjacent traders and businesses,
  regarding programming and site
  activities shall be maintained from
  pre-start to completion

Responsibility
Considerate Constructors will ensure that
  all site personnel, specialist subcontractors,
  drivers and any other persons working on
  the site understand and implement the
  obligations of this Code, and will monitor
  their compliance with it

Respect
Respectable and safe standards of dress
  shall be maintained at all times.
  Lewd or derogatory behaviour and
  language should not be tolerated,
  under threat of severe disciplinary
  action. Pride in the management
  and appearance of the site and the
  surrounding environment is to be
  shown at all times. Operatives shall be
  instructed in dealing with the
  general public

Cleanliness
The working site is to be kept clean and
  in goodorder at all times. Temporary
  safety barriers, lights and warning signs
  are to be maintained in a clean and
  safe condition. Surplus materials and
  rubbish shall not be allowed to
  accumulate on the site or spill over on
  to the surrounding environment.
  Dust from construction operations
  shall be kept to a minimum

Safety
Construction operations and site vehicle
  movements are to be carried out
  with care and consideration for the
  safety of the general public, traders,
  shoppers as well as site personnel.
  No building activity shall be a
  security risk to others

Accountability
CCS posters are to be displayed around
  the site, giving names and telephone
  numbers of staff who can be
  contacted in response to issues
  raised by the general public, traders,
  shoppers and others affected by
  the site operations

Source: http://www.considerateconstructorsscheme.org.uk/htm-whatwedo/practice.htm (accessed 05/08/05).

regional coordinators to establish clear reporting procedures (so as to enable central records updates) was seen as a vital communications step.

Clearly, the daily pressure of managing construction projects could result in the CCS scheme not being considered a top priority. From a communications perspective, it was therefore vital that HBG's Chief Executive acknowledged this and appealed for his workforce to whole-heartedly adopt the scheme:

> The need for responsible community relations is not intended as an additional burden to your workload, but it is aimed at reducing external complaints and distractions that unhappy neighbours can cause you. The investment in a little time for the local community early on will pay long-term dividends in the overall construction process. Good citizenship reflects a good company. It also makes good business sense. Our behaviour on site reflects upon our client. Respect for people and a commitment to nurture understanding between our neighbours and our staff have to be a major priority in all our construction processes.

The approach adopted by the HBG document provided an honest and transparent appraisal of where the company was and the challenges it faced in improving its CSR policies. It was recognised that HBG was performing significantly above the industry average, but challenged staff to maintain their success in the annual CCS Awards. The report recognised three areas for improvement, however, namely in being a 'good neighbour' promoting cleanliness and demonstrating accountability. Table 6.3 outlines some of the guidance communicated to staff in improving performance in these three areas.

Post project community reviews

As a means to assess a local community's perception of HBG's CCS at individual project level, the PR team at HBG have undertaken post-project reviews. Where a construction phase of a project exceeds 18 months, the PR team send out a short questionnaire to a selection of local residents and businesses that may have been affected by project operations. As a means to encourage completion and return of the questionnaire, recipients are entered into a prize draw. The questionnaires are returned direct to the PR team as this provides them with a secondary source of data. This data is fed back to the client, the project team and regional management teams to encourage further improvement in future activities.

Communicating the CSR policy: internal and external communication processes

Regular, effective communication across the organisation is seen as an important part of the HBG business strategy and they operate a number of different methods to achieve this (HBG, 2004b). The company has gone to

Table 6.3 Communicating best practice guidance for compliance with the CCS scheme

Good neighbours	Cleanliness	Accountability
Redistribute to all sites 'A practical guide to best practice for site-based responsible community relations' Refer sites to the Intranet (Directories/Press and Publicity/Guidance/ Community Relations) and site newsletter for further help and templates Liaise with Graphics and Public relations department to procedure more frequent and better quality templates	Remind sites of our Environmental policy to minimise waste, segregate where possible and contain all waste materials Refer sites to the Environmental Management System (EMS) document on the Intranet Ensure all subcontractors are responsible for clearing up their work continually – but particularly once a task is completed Posters for canteens and cabins will be produced and circulated	Ensure all newsletters contain relevant site name(s) and contact details for receiving complaints Liaise with Graphics/Public relations Department for bespoke posters/signage. Ensure the site security cabin/office gives sufficient information for the public to have access to a complaints procedure

Source: HBG (2004a).

great efforts to ensure effective communication both within the organisation and to external agencies.

Internal communication processes

HBG has developed a range of mutually supportive and reinforcing communication channels for ensuring that staff are well informed with regards to their CSR policy. For example, an in-house magazine of HBG (People) is distributed to 2,100 employees, company pensioners and sponsored students. The magazine contains news about staff and company activities. In addition, a further Group magazine (Bam World) is published five times a year and distributed to all staff with global group news. Most staff in all regions attend annual company 'road shows' which are supplemented by each business unit communicating with staff through local events and discipline-specific gatherings. Such events can be used to capture employee opinions on corporate policies as well as disseminating news and information throughout the organisation. The company also utilises an intranet featuring company, industry and national news. There is also a fully targeted 'HBG Newsroom' email established for important corporate use. Thus, the approach of HBG to ensuring effective internal communications does not give primacy to any particular communication mechanism. Rather, any channel germane to getting information efficiently and effectively to staff is utilised. By ensuring that different methods are used in concert this increases the likelihood of employees being exposed to information and hence, to them acting upon it.

External communication processes

Although internal staff communication mechanisms are important, another very important element arises at the interface between the site and the local community. HBG recognises that an informed and regularly updated local community will be far more understanding and tolerant of the disruption that inevitably arises from construction work (HBG, 2001). HBG's guide for site-based managers details practical guidance for ensuring best practice community relations. Each new project has a single point of contact responsible for community relations, and they are clearly nominated as such on the site organisation plan. This role involves instilling all those involved in the project, with a sense of commitment to behaving in a considerate and respectful attitude towards the public. They must use all practicable means to communicate with the neighbouring residential, commercial, academic and local communities and to act as the main conduit for the receipt, processing and reporting of complaints. Initiatives to facilitate this community liaison role include residents meetings, open evenings, school liaison and media communications. Another important external communications role is in ensuring that the company's supply chain supports and works in a way which is aligned with the company's needs. HBG have embarked on a programme of disseminating their CSR aspirations throughout their supply chain to ensure compliance with its policy. Measures to ensure this include strict pre-qualification checks of its subcontractors and suppliers. Finally, HBG closely aligns its CSR policies and programmes with those of its clients by continuous dialogue to ensure compatibility of policies, processes and delivery.

Conclusions and lessons learned

This case study has provided an excellent example of how a large and diverse company can develop robust corporate policies and practices and communicate them effectively to both its workforce and to important external stakeholders. The success of HBG in implementing this policy can be seen to stem from its high level commitment to the endeavour (i.e. the endorsement and hands-on management by company directors), its willingness to put the company up to the scrutiny of an externally validated best practice standard (i.e. the CCS metrics), its integrated internal communications processes (i.e. the way in which different channels are combined to ensure the effective communication of company policy) and the way in which external stakeholders are engaged (i.e. through leveraging the support of the supply chain and liaising with the local community). The result of this integrated and highly cogent strategy is that the company is far more likely to present a coherent and responsible image to both its internal and external stakeholders. These themes are explored further in the following chapter.

Chapter 7

Corporate communication

Introduction

The construction industry needs to establish ways to communicate more effectively with the outside world (Preece *et al.*, 1998: v). Corporate communication refers to the ways in which an organisation relates to its external environment, including its clients, competitors and supply chain. Whereas Chapter 6 focused on how an organisation can develop effective internal and external communication systems, the focus of this chapter shifts to consider the requirements of communicating a company's corporate image to the outside world, which is particularly important in portraying a positive image to both its client base and to other members of its supply chain. This chapter explores the topic of corporate communication from two perspectives. First, the topic is examined from a traditional perspective by discussing key aspects of corporate communication such as corporate identity and image and how the process of re-branding can be managed by construction companies. Later, it highlights the important role that individual projects can play in providing construction businesses with opportunities to communicate their vision, beliefs, identity and brand. Thus, from a project manager's perspective, this chapter will help to reinforce the wider responsibilities that this professional will have to both his/her organisation and to the client in projecting the corporate values of a project and safeguarding the image of the parties involved.

Defining corporate communication

Defining 'corporate communication' is difficult because it is a nebulous and all-encompassing term which will mean dissimilar things to different organisations. Academics and scholars place different emphases on certain aspects of corporate communication to practitioners and so there remains a definite tension between theory and practice (van-Riel, 1995). The common dominator, however, is that most definitions refer to the importance of communicating a positive image for the entity concerned. In this regard, Jackson (1987) provides a broad overarching definition of corporate communication that is useful in that it can encompass all the activities that could be considered to fall under this term: 'The total communication activity generated by a company to achieve its planned objectives'. This broad definition encompasses activities such as advertising, public relations, investor relations and corporate design, all of which will be covered in this chapter in relation to construction companies.

Atkinson (2002) notes that corporate communication has its origins in the actions of firms of the late nineteenth and early twentieth century, who employed advertising and public relations managers so as to provide an image for their companies and their products. In particular, Atkinson refers to the need for such organisations to demonstrate that 'Victorian' values were prevalent, whereby paternalistic corporations could be seen to link the business to family and community. Frankental (2001) also comments on the origins of Corporate Social Responsibility (CSR) but argues that the genuine philanthropic actions (taken by the likes of Joseph Rowntree) are not consistent with the twenty-first-century concept of CSR. He argues that organisations today are motivated by a desire for eventual return, a more compliant workforce and more amenable customers.

In exploring the concept of corporate communication in more detail, van-Riel (1995) describes three main forms of the function: marketing communication; organisational communication and management communication. These themes encompass aspects of both internal and external communication, in that every employee should be responsible for communicating the organisational 'ethos', 'visions' and 'values' to both colleagues or customers alike. This will act as a reinforcement of a single corporate identity, which van-Riel (ibid.) refers to as the 'orchestration into a coherent whole'. In this respect, an organisational communications policy should be derived from a set of common stand points that are themselves derived from the corporate strategy identity and desired image of the firm. Moreover, they should appeal, and be recognisable, to all relevant organisational stakeholders as is explored later.

Stakeholder theory

Stakeholders can be defined as any individual, group or organisation with a stake in the firm with which we are concerned. That means that they are affected in some way by the achievements (or the lack of achievements) of the organisation. These need not be people or groups who necessarily stand to gain from the success of firm. Consider, for example, a local action group objecting to planning permission being granted for a new housing development. Although this group stand to benefit from the organisation's misfortune should they not be granted permission to construct the development, they could still be considered stakeholders to the organisation as they are directly effected by their activities and indeed, have the potential to affect their activities. It is still incumbent upon the organisation to consider their corporate relations with this group, even though their motives and aspirations may be dichotomously opposed.

Stakeholders in the construction industry can include clients (private and public sector), government (local and national), suppliers, subcontractors and anyone else affected by the built environment. However, for practical reasons it is incumbent upon most organisations to define who their primary stakeholders are in order that they can focus their efforts and attention on communicating a positive image to this group. Primary stakeholders can be defined as those with a close proximity to the core business of the organisation (i.e. those with a direct and economic impact upon it – see Preece *et al.*, 1998). Groups with an adversarial relationship with the organisation can be considered secondary stakeholders in this context. Consideration should also be paid to the type of stake that each stakeholder has in the organisation. For most of the primary stakeholders, this will be an economic influence, but others may be purely technological, social or political. Once identified, the organisation should ensure that its approach and philosophy aligns with those of its stakeholders in such a way that a shared value system emerges. This will inevitably require proactive steps on behalf of the organisation to ensure that stakeholders understand their corporate philosophy and the benefits that accrue to them from such a business approach. Failure to manage stakeholder expectations effectively will inevitably result in external environmental pressures on the firm which may ultimately impinge on the performance of the business. The importance of identifying relevant primary and secondary stakeholders will be further developed within the discourse around corporate and social responsibility later in this chapter.

Principles of mass communication

Until this chapter, this book has largely been focused on specific communication strategies between individuals, teams and groups within projects and organisations. Corporate communication, on the other hand, demands that a much broader perspective on communication is considered. The nature of general communication differs from that of specific communication in at least one important aspect – the nature of the encoding process. Initially, there may be the perception that general communication is less demanding than specific communication. It seems reasonable to assume that the 'message' to be conveyed does not need to be as precise in its content as one that is aimed at an individual or specific group. However, the diversity of recipient characteristics (i.e. their language, culture, values, etc.) increases with the increasing size of that group, unless its membership is homogenous. In essence therefore, the challenge with mass communication is that those responsible for corporate communication have to consider how many individuals respond to the message and the medium though which it is conveyed. Rather than struggle with such a level of complexity, most mass media communicators (such as advertising executives) target aspects of individuals that can be regarded as (largely) generic, and which can be regarded as being groupthink by default (see Chapter 5 for a discussion around the nature of groupthink within a communication context). A good example in popular culture are car advertisements, which today tend to focus on manipulating pre-existing attitudes and social values through the image of the product rather than on selling the technical specification of the vehicle *per se*.

The role of the publicist

Corporate communication professionals (or 'publicists') may be employed within a range of departments within an organisation, for example customer relations, marketing, business development or public affairs. In this regard, Dolphin (2002) suggests that what used to trade under the name of Public Relations (PR) is now variously defined as corporate affairs, corporate communications and/or public affairs. Correspondingly, there are different job titles for the executives holding such positions. Dolphin's review of PR directors within 20 UK organisations revealed that in all but one case, none had a marketing background.

The role of the publicist is often associated with professionals working in the media. Professional publicists are seen to 'smooth out' the problems of high profile people or to 'spin' or make the most of potential news stories on behalf of their clients. Indeed, some politicians and military strategists have created a new language to communicate bad news in a damage limitation mode. A good example is the description of casualities within war zones. Terms such as 'collateral damage' (civilian casualties) and 'friendly fire' (killing your allies) could be considered unethical despite their appeal. Within a construction company context, the role of the publicist could be to maximise the benefits of company achievements (such as the early completion of a high-profile project) or to limit the damage caused by an unsuccessful event (such as delays to a high profile project – see the case study in Chapter 2.

Brand, identity and image

Despite having roots in the Victorian era, corporate communication is considered a relatively new and evolving discipline. It is therefore not surprising to find conflicting evidence when attempting to define terms such as brand, identity and image. Van-Riel (1995) provides an excellent literature critique and provides definitions that allow practitioners to understand the plethora of concepts that are often associated with these terms. He provides an in-depth analysis of corporate identity and corporate image and notes the confusion that exists amongst the professionals who are employed to manage corporate identity. This analysis reveals that many practitioners consider corporate identity and image to be synonymous. Indeed van-Riel's review of the literature on this matter (ibid.) concluded that no consistent meaning of the term 'image' or of the ideal method of measuring it could be found. He argues that in spite of their relationship, the terms image and identity are fundamentally different. Table 7.1 shows two distinct definitions used by van-Riel in his textbook, together with extracts from the book that emphasise the differences between corporate identity and image.

Balmer and Gray (2003) also contrast 'identity' with 'brand' by arguing that an organisation's identity is a prerequisite to the establishment of a corporate brand. It is also important to distinguish between product brands (of primary importance to consumers) corporate brands (concerning a myriad

Table 7.1 Corporate identity and corporate image

Corporate identity	Corporate image
The self-portrayal of an organisation that is, the cues or signals which it offers via its behaviour, communication and symbolism (van Rekom et al., 1991)	The set of meanings by which an object is known and through which people describe, remember and relate to it. That is, the net results of the inter-action of a person's beliefs, ideas, feelings and impressions about an object (Dowling, 1986)
• The way in which a company presents itself to its target groups • The manifestation of a bundle of characteristics, which form a kind of shell around an organisation, displaying its personality • Undertaken through the use of symbols, communication and behaviour. These media together constitute the corporate identity mix • Can be considered as a kind of adhesive that bonds both internal and external target groups • They are the concrete forms into which the company's personality crystallizes	• The picture that people have of a company • Reflects the identity of an organisation • Can be product image, brand image, company image, industry image, user image • Personal impressions, interpersonal communication and mass media communication combine to produce a mixture of real and parallel impressions, the totality of which forms the image • 'Impression management' is a way of creating or protecting an image among members of target groups. It is the company's policy of presenting itself in such a way to evoke a favourable (image) or to avoid an unfavourable picture (van Raaij, 1986)

Source: Cf. van-Riel (1995).

of stakeholder groups) and corporate identity (ibid.). A corporate brand is a valuable resource that can provide sustainable competitive advantage if it is characterised by value, rarity, durability and imperfect substitutability. Table 7.2 summarises the differences between corporate brand, product brand and corporate identity and provides some examples taken from the construction industry.

Once established, a brand can have a life of its own: it can be bought, borrowed, sold and in certain circumstances, shared amongst a variety of organisations. A good recent example this within the UK construction industry concerns the corporate image and identity of the Laing O'Rourke organisation. This company was formed after the sale of Laing Construction to former subcontractor O'Rourke in 2002. Here, the Laing

Table 7.2 Corporate brands, product brands and corporate identity

Term	Defining characteristics	Construction examples
Corporate brand	Chief executive level responsibility involving most departments and all personnel and therefore an important element of company strategy. Corporate brand values tend to be grounded in the values and affinities of company founders, owners, management and personnel. They are clearly articulated, concise, well defined and distinct and broadly constant over time	A chief executive making a speech at a company conference in order to communicate his/her ethos to staff and external stakeholders
Product brand	Involves brand managers with specific concern for marketing. Focuses on consumers and tends to involve short-term brand gestation. Product brand values tend to be contrived and are the product of the skills of invention held by marketing and advertising professionals	Construction component manufacturers develop brand identities for their individual products and contractors for their specialist sub-divisions
Corporate identity	An organisation's identity encompasses a bundle of values that are derived from a federation of subcultures, which are found within and outside the organisation. They are amorphous and evolve continually. This mix of values gives an organisation its distinctiveness	Many projects co-locate team members to reinforce a single project identity. This transcends the boundaries of the organisations involved as they become defined by an amalgam of the cultures and approaches of those involved

Source: Cf. Balmer and Gray (2003).

brand was maintained in the title of the new company because of the stature and standing of the company name within the industry.

Branding can be used to widen an organisation's exposure, perhaps to stakeholders that are very difficult to reach. Most construction firms provide a range of branded calendars, pens and coffee mugs etc. One UK contractor has taken a novel approach in this direction and have a 'kids zone' (Kieran) on their website (Kier, 2005). This offers children an inter-active experience including a story book, colouring and a safety quiz with certificate. In addition Kier's shop sells a wide range of branded goods for children that can be ordered through Kier's public relations department. This appears to be a unique idea and has the advantage of fulfilling a corporate citizenship role whilst offering exposure in other ways (future recruitment).

Company logos

In addition to names and titles, company logos are extremely powerful ways of conveying a company image, standing and reputation. A company logo may not always display a name (consider for example the sporting goods manufactures Nike and Adidas which are better known by their 'tick' and three stripes respectively), but will nevertheless communicate their name to the receiver rapidly and effectively. People instinctively learn logos from a young age and where managed particularly effectively, product names change to reflect the brand name (e.g. many people refer to excavators as 'JCBs' rather than their technical names). The most highly valued logos tend to be registered trademarks which are jealously guarded by their owners as they are deeply associated with a product.

Logos do not become instantly recognisable, but over time, become associated with a particular brand or image. A good example of the evolution and development of a company logo is the leading contractor Bovis Lendlease, who had used its company name on site boards and vehicles for several decades (see Cooper, 2000). By the early 1970's Bovis were looking for a group symbol as part of a drive for a clear corporate identity. In the preceding years it had acquired over twenty construction and property companies. Wolf Olins, a design consultancy, was commissioned to establish a new identity and the famous 'hummingbird' was selected to represent the values of the group's precision and industriousness. It was claimed that the new look gave the company a huge promotional edge by differentiating it from competitors whose logos were characterized by heavy impersonal identities (Wolf Olins, 2004) or as Williams (2004) observes, big and anonymous symbols. Cooper (ibid.) notes that Bovis executives were to learn of the ornithological appropriateness of the bird, after, the selection process; the hummingbird's nest being a perfectly formed building structure delivered just in time for its eggs! It can be seen, therefore, that corporate identity, as embodied by a company logo, is a vital strategic tool that plays a crucial role in a company's success (Preece *et al.*, 1998: 16).

Re-branding an organisation

Engaging a business in a re-branding exercise will normally involve changing a logo and with it, the corporate identity of an organisation. When a re-branding exercise is attempted care should be taken to avoid losing the association with a particular name or branding that the logo may convey. A good example of where this has been compromised is the 1997 British Airways £60 million re-launch which involved replacing the Union flag on its tailfins with 28 world image designs. However, the images were not well received and even during the launch, the former Prime Minister, Margaret Thatcher, famously draped her handkerchief over a model with the new design. By 2001, the decision was taken to restore the Union flag to the tailfins, apparently so as to reinforce the company's 'Britishness' (*The Telegraph*, 2001).

The construction industry has seen both successful and unsuccessful attempts to re-brand. Some well-known positive examples include contractor Tilbury Douglas, who changed its name to Interserve after their move from the 'construction' listing to 'service industries' on the London stock exchange, and the Tarmac organisation which, following de-merger, created the Carillion brand with an associated new logo. Both organisations have arguably benefited from the new image that such a name and logo change conveys. However, such a change is not always as successful. In 1999, contractor Alfred McAlpine embarked on an exercise that saw them dropping 'Alfred' from their trading name and using a new logo that displayed 'McAlpine' only. When the new logo was introduced in 2003, Sir Robert McAlpine claimed that it broke a historical agreement by both firms that neither would rename themselves 'McAlpine' so as to avoid confusion between the brands. Despite Alfred McAlpine successfully securing a patent licence for the name and expressing a willingness to share the trademark, it was banned from using the name as a trading name following legal action (*Construction News*, 2004a,b).

Managing the corporate image of a construction organisation

The negative image of the construction industry has arguably been propagated by the media who appear to use the sector as a focus for consumer concerns around poor workmanship and questionable products. Television programmes abound which highlight the work of so-called 'cowboy builders' or which uncover problems in the quality of buildings and structures. In the

1980s, one such programme condemned the quality control process in timber-frame house building construction. This programme severely affected consumer confidence in this construction technique and resulted in a severe downturn in the number of houses built using this technology. This example emphasises the linkage between corporate identity and corporate image and moreover, the power of the media in shaping the public perception of the industry. It also emphasises the importance of providing construction professionals with guidance on managing the media. Indeed, this issue is so serious that some construction firms regularly send their key managers on courses that train them to handle media interviews, as the ability to steer the interview is seen as being as critical as the need to talk with integrity and conviction (*Construction News*, 2002a). Others employ public relations experts to manage the reputation of firms and the relationship with the media (see *Building*, 2001c). However, it is incumbent upon all involved in the construction process to be aware of their responsibilities with regards to protecting the image of their organisation. Construction firms can project a positive corporate image through both their corporate activities and their projects. However, both require careful management if a desirable (and consistent) image is to be portrayed as is explored later.

Managing corporate image at the organisational level

In addition to logos and branding, the main way in which large organisations convey a corporate picture of their performance is via statements of their mission, values and performance. This is frequently portrayed through their corporate annual reports which often contain information about an organisation's history, business objectives and corporate philosophy. They are a key vehicle for communication between a firm's management and its stakeholders and are a primary source of financial and operating information about the firm. Of all publicly produced documents, annual reports are the most widely distributed (Campbell, 2000).

The focus of a company's strategy is usually reflected in its mission statement. This states company goals with regards to its clients, technologies, concern for the environment and commitment to its employees. Edum-Fotwe *et al.* (1996) undertook a study of mission statements in construction contractors' annual reports. They found that contractor's strategies embody five main factors, namely: services and markets, location and geographical spread, economic survival and profitability, philosophy and values (aspirations) and self-concept (strengths and weaknesses). Their findings suggest that construction companies tend to be more concerned with their market position and stability than their public image. Considering that annual reports are believed by many clients to provide detail of a firm's philosophy and standing (Preece *et al.*, 1998: 51), it is vital that they are well designed, easy to understand and reflective of the company's approach and ethos.

Preece *et al.* (1998: 76) offer a simple checklist of key points to consider for construction firms when developing their accounts strategy. Key elements to consider from this list include the need for clear and concise forward-looking statements from senior managers supported with biographies of their achievements, feature sections to support the report which showcase the success and competence of the company, financial summaries which show how the company is doing over a defined period and an assessment of the overall state of the market within which the company operates. There is also usually a general statement of how the company intends to react to threats and opportunities.

The language used in annual reports can be examined by content/discourse analysis to determine the level of disclosure (Beattie *et al.*, 2002). Within the construction sector, Green (2002) examined 27 selected annual reports of top construction organisations. This analysis noted the occurrence of certain buzz words or phrases such as *'deeper relationships with our customers'* and *'in partnership with our customers'* that appear to be de-rigueur within the reports examined. He noted the regular occurrence of the word 'customer' rather than 'client' and concluded that this is partially due to pressure from the London Stock Exchange and its fund holders who have pressed construction firms to shift from a 'production' to a service-based business. He argues that by changing their language they appear to be seeking to instil a cultural shift throughout their businesses in terms of how the outside world perceives them (corporate image).

Moves to make listed companies more transparent are underway. In April 2005 the Operating and Financial Reviews (OFR) regulations are to come into force within the United Kingdom (DTI, 2004). The OFR is intended to improve the quality, utility and relevance of information provided by quoted companies, helping shareholders get a better understanding of a quoted company's business and future prospects. The mandatory requirements will mean that companies will be required to provide an overview of their strategy, past performance and future prospects, as well as information about its ethical stance. It is likely that this information will be collated within a separate section in annual reports. It would appear that the UK's top construction companies have already made progress towards greater transparency, typically by publishing safety, environmental and social reports that provide much more information than is usually incorporated in annual 'profit and loss' accounts.

Accessing information such as annual reports and supplementary documents is now typically done through accessing an organisation's website. Many construction companies have extensive information on their sites, where documents can be downloaded. Research undertaken by Cox and Preece (2000) and Preece *et al.* (2001) led to the production of the first league table showing the most effective websites of the top fifty UK contractors. The study involved establishing a way of assessing websites and included a focus group of construction professionals representing a cross section of the industry to identify the assessment mechanism to be applied. The research also tested ease of access through commercial search engines and the response times of contractors to requests for information. The researchers concluded that a website not only needs to contain the quality and depth of information client teams are looking for, it also needs to be usable and to have a fully functional back office system in place. It would appear that the majority of the UK top construction companies understand that their website acts as a window for clients, graduates and all other stakeholders who wish to learn more about the company's operations and its corporate principles.

Managing corporate image through projects

The role of the project manager in corporate communication may seem fairly marginal on initial consideration. However, as was alluded to earlier in this chapter, projects are the conduits through which a construction company conveys its expertise, quality and achievements to the outside world. The success or otherwise of high profile projects can have a lasting legacy for those involved and can even define the future direction and standing of an organisation. A disastrous project in particular can have long-term implications for the future marketability of the firm. These principles apply as much for small firms operating within predominantly local markets (where word-of-mouth recommendations often yield new business opportunities) as it does for major firms constructing high profile developments. Thus, an effective strategy for ensuring that a positive corporate image is communicated at the project level is a prerequisite for the success of all construction businesses.

Considering the fragmented nature of their delivery and the multi-layered supply chain, many construction projects will 'unintentionally' communicate different messages to many stakeholders. Indeed, given the multi-organisational nature of construction projects, it is difficult to find examples of where one project has a unified corporate message or ethos that has been intentionally communicated to external parties (see the case study at the end of this chapter). The 'brand value' may take several years to emerge and can be an unexpected rather than a planned benefit. The reasons for this are rooted in the separation of the activities comprising the construction effort and inherent subjectivity in judging the merits (or otherwise) of construction projects. The visual stimuli received by a person observing the aesthetics or structure, for example, mean that the merits of a building are in 'the eye of the beholder' and are partially a product of that individual's values and beliefs. Perception can be influenced through the efficacy of the design process, but cognition by those stakeholders who were not party to the design process cannot not be controlled after the design has been implemented and the building constructed. However, although the cognition stimuli that a person observing the aesthetics reacts to cannot be controlled, the appearance of the site (particularly around its perimeter), sends direct messages (through visual perception) to all sighted project stakeholders. Moreover, given the number of deaths and injuries within the construction industry, safety and welfare continues to be an extremely important lens through which the public builds its image of the industry and individual projects. In this respect, Preece *et al.* (1998) note the importance of gaining support from the local press and media who can help spread safety awareness within local schools and the proximate community.

As was discussed earlier in this chapter, it has become generally accepted that *image* is the picture of an organisation as perceived by target groups, whilst *identity* is associated with the way in which a company presents itself to its target groups (van-Riel, 1995). Considering the multitude of stakeholder interests found within a typical construction project, the identification of 'target groups' may be a more difficult process for construction companies than for companies operating in other industries. Clearly, the client is the prime target in that employment on any future projects is likely to be dependent on current performance. Projects also present a window for potential new clients to view organisational performance. The multi-organisational nature of construction projects also suggests that marketing opportunities may also exist *within* the project supply chain itself. For example, an engineering consultant may 'target' project design colleagues in the belief that this may provide opportunities for future commissions. Indeed, it could be considered that all of the project stakeholders are 'target' groups and they spend time both presenting their own identity and making judgements about others' corporate image. Thus, it is little wonder that the construction industry struggles to present a common identity through its projects.

The most obvious signs of corporate identity in a project are likely to be those based on the use of a company name or logo on a project board, hoarding or mechanical plant. Knowledgeable and repeat-build clients often lead the way in this respect with some (such as supermarkets) running site branding initiatives. For example, new-build Tesco projects have identical blue hoarding and the construction workforce wear overalls branded with both the Tesco logo and employees' company name (*Building*, 1997). Other ways in which a positive corporate image can be portrayed through the project is to work responsibly in connection with, and be sympathetic to, stakeholders' needs. As was discussed in Chapter 5, the Considerate Constructors Scheme was established in 1997 to promote responsible construction work. Contractors who register a project agree to abide by an eight-point code of practice, committing them to be considerate and good neighbours, as well as clean, respectful, safe, environmentally conscious, responsible and accountable (Considerate Constructors Scheme, 2004).

Some construction workers on high profile projects have been exhorted to smarten up their appearance (*Construction News*, 2002b) as the Considerate Constructors Scheme even demands an 'appropriate' dress code as an important part of demonstrating company and industry image. Such initiatives encourage those involved in the project to adhere to a way of working which belies the industry's traditional stereotype and which recognises the impact the construction work has on those within close proximity of the industry's projects.

One problem that project managers face when trying to convey a positive corporate image of their firm through the projects that they manage is the unpredictability of construction endeavours relative to those undertaken in less dynamic and more easily controlled industries. Prestigious projects with a high level of public interest are most exposed and it is often difficult for the public to determine who the guilty party is when a crisis occurs or when a project does not go according to plan. A good example of this is embodied within the case of the Millennium Bridge project, a footbridge over the River Thames in London. The innovative structural design resulted in the footbridge 'wobbling' under human traffic. The subsequent closure of the bridge increased media attention. The design engineer took the decision to be open and honest about the bridge's failings and why it had to be closed (Ryan, 2001). Although a well-published case, those involved seem to have benefited from the publicity associated with the bridge and the eventual impressive, fully functioning product which has emerged. This aspect is discussed by Miller and Lessard (2000) who conducted an in-depth analysis of sixty Large Engineering Projects (LEPs). They refer to the benefits derived from the Eurotunnel rail service and argue that despite its initial harsh treatment from the media, the project was well designed to meet real and socially valuable objectives. Thus, as with the bridge project, the media and public's evaluation of the project eventually rebounded.

The management of risk is an important factor in any large engineering project. In mega-projects such as the new Terminal 5 building at Heathrow, the client (British Airport Authority) was clearly aware that the eyes of the world were watching. Following one of the longest public enquiries in British planning history (1995–1999) the project started in 2002 with a completion date set for spring 2008. The British Airport Authorities (BAA) gained approval for the new terminal and associated facilities only after agreeing to accept stringent planning conditions particularly regarding air and noise pollution levels. With this in mind, they established the T5 Agreement, the contract for the project that involves BAA accepting that they carry all of the risk in the project. As the project's commercial director notes 'the client is always accountable in the end, on cost, time and health and safety-everything. If we fail, the impact on our reputation and relations with shareholders and the City would not be worth contemplating' (Riley, 2004). Indeed, the whole project could be considered an exercise in reputation management and BAA and its suppliers have set new standards in logistics management, occupational health and safety and environmental management. However, despite these innovations, the project itself is under constant scrutiny from its objectors who argue that an increase in air travel results in further global pollution and will no doubt lead to an increase in local pollution from additional car journeys to and from the new terminal. It is difficult to offer a robust and valid response to such criticism and to some extent this is indicative of the challenge set for the construction industry and in particular its publicists who may need to positively promote the advantages of construction and infrastructure development.

Coping with crises

Inevitably, construction projects can face crisis events, the cause of which may be outside of the direct control of the key participants. It is a fact that in modern society, the media are playing an increasingly significant role in shaping the public perception and reaction to such crises. Thus, managers need to develop the skills to deal with the media effectively, and to manage their output which will have knock-on effects in terms of the public perception of the handling of a crisis event. In the United States, the management of crises in construction is arguably more advanced than in the UK. There are even management consultancies who offer a Crisis Management Planning Package designed specifically for the construction industry (Reid, 2000). It is clearly beneficial if project managers are trained in coping with crisis situations in order that they can mitigate the damage to corporate image through their immediate management of the situation.

There are a number of general rules in controlling media intrusion during a crisis situation. By putting these in place, the project manager can focus attention on controlling media interaction:

- *Control of where the media may go* – for crises with considerable national or international interest, it is essential that media control appears planned, with a policy for stating where the media can and cannot go. A more controversial technique is to allow limited numbers of media representatives to access the crisis site, but for the material collected by these representatives to be shared amongst the media groups.
- *Shaping answers to media questions* – by rephrasing media questions the conversation can be led so that the interviewee makes the points that they want to make. For example '...Although I can't answer that now, I can state that we...'.
- *Sound bites* – restricting information to between 10 and 30 seconds are more likely to be broadcast rather than long explanations of particular situations.
- *Deal in facts* – focus on what is seen as correct, true and factual rather than assumptions.

- *Remain calm and speak as a person, not as a spokesperson* – by personalising statements and comments, whilst appearing calm and open, this gives the impression of being concerned and in control of crisis situations.
- *Adopt a helpful, non-blaming position* – by avoiding saying 'no comment' (which infers that information is being concealed), avoiding being misinterpreted by explaining why new information contradicts past information, avoiding speculative statements and avoiding assigning blame to people or organisations, this will portray a better image to the media, and will avoid clashes.

The principle of these techniques is to allow the project manager more time to draw upon the resources of the wider organisation in managing the reaction to the crisis event. It is essential that the project manager retains *an outward focus*, as assigning blame or defensive attitudes present a negative image. Presenting positive actions to help those affected by a crisis is interpreted as having an open and caring attitude. Thus managers must assess who the stakeholders affected by the crisis are and focus attention on helping them recover.

Managing the corporate image of the construction industry

Considering that the construction industry is of considerable economic and strategic importance across the United Kingdom, its role in society is often ignored or taken for granted (Dickson, 2003 cited in Pearce, 2003). It is important that the sector attempts to foster a more positive image, highlighting its contribution to the economic well-being and social regeneration of the country. It is particularly important to improve the corporate image of the sector if skills shortages are to be avoided and the industry is to attract high performing individuals to its professional, managerial and craft occupations. Central to improving the industry's image is the promotion of its career pathways and conveying a sense of it being a progressive, rapidly developing sector.

Promoting construction careers

In the past decade, a recurring theme within the industry trade press has been the widespread reports of the difficulties faced by construction companies in recruiting skilled labour throughout the UK sector. In particular, maintaining a steady flow of school leavers into craft apprenticeships has proven problematic (see Agapiou, 2003; Keel *et al.*, 2004). A survey of four hundred, 15–17-year olds commissioned by *Building* (1999) revealed that only 49 per cent viewed the construction industry favourably and only 27 per cent

said that working in construction appealed. A MORI research study commissioned by the CITB in 1998 provided an analysis of over four thousand
11–16-year-old secondary school pupils and also revealed a high level of
ambivalence towards the construction industry (CITB, 1998). Nearly three-
quarters of those surveyed mentioned at least one 'negative' attribute, that
it is dangerous, dirty or badly paid. These findings should not be so surprising given that many young people see construction as unsafe, poorly
paid and as environmentally unsound (see Bale, 2001).

As a means to combat the problems noted above the CITB embarked on
a billboard and advert campaign aimed at re-branding careers in construction.
This initiative was part of an ongoing programme which sought to present
a positive image of careers in construction. The campaign was targeted at
14–19-year olds using full page advertisements in national press and youth
magazines showing various young people who had taken up careers in construction. The adverts were designed to appeal to young people's interests
in money, music, travel, socialising and sport. In addition postcards were
placed in cinemas during periods when popular teenage films were being
screened. They also focused on encouraging more women and ethnic
minorities into the industry with the intention of creating heroes in construction for young people to look up to. Other initiatives/associations that
promote a more diverse workforce in construction include:

- *National Construction Week (NCW)*: This is a UKwide not-for-profit
  campaign which gives young people the opportunity to experience the
  wide range of opportunities available in the modern construction
  industry (see www.ncw.org.uk).
- *Change the Face of Construction*: Change the face of construction is
  an independent project dedicated to encouraging greater diversity
  across all sectors of the construction industry. The aim is to improve
  industry performance by attracting and keeping more of the right
  women and men, through better recruitment and training, working
  conditions, career development and communication (see www.change-
  construction.org).
- *National Association of Women in Construction (NAWIC)*: Women
  currently make up just 10 per cent of the UK construction workforce.
  The NAWIC aims not only to redress this balance, but also to raise the
  profile of those professional women already working in the industry,
  encouraging best practice and providing mutual support (see
  www.nawic.co.uk).
- *Women and Manual Trades (WAMT)*: WAMT is the national organisation for tradeswomen and women training in the trades. WAMT aim
  to represent and support women working and training in skilled
  manual and craft occupations in industries where women are underrepresented (see www.wamt.org).

- *The Association of Women into Property (WiP)*: WiP is a dynamic forum for the professional development of women in the property and construction industry which aims to enhance business opportunities, exchange views, network and gain knowledge. It is a unique organisation with a multidisciplined membership drawn from across the property and construction industry. The organisation seeks to promote the role of women in a wider business community and to encourage women to take up and develop their careers (see www.wip.propertymall.com).

Such initiatives are gradually promoting the appeal of the sector to underrepresented groups.

Promoting construction as a 'progressive' industry

Since the publication of the *Rethinking Construction* report (Egan, 1998) the construction sector has been subjected to a plethora of 'Eganised' initiatives which show no sign of relenting (Murray and Langford, 2003). The calls for radical improvements to the industry's performance have driven the industry towards a culture of continuous improvement. Indeed, as Murray (2003a) notes, the *Rethinking Construction* report has spawned a new language and narrative within the industry. 'Egan' itself is understood to be both a noun (person) and vowel (to improve the process of construction, to remove waste, to increase efficiency, to rethink processes etc.). Although some are partaking in the improvement process, it is clear that the vision for a world class construction industry has been ignored by too many. In 2001, the Construction Best Practice Programme (CBPP) estimated that it had only reached 9 per cent of the working population in the industry through its programme of seminars and workshops (National Audit Office, 2001). A more recent survey of 1,340 workers (public and private clients, contractors, consultants) employed in the construction industry also revealed dissemination problems. The survey, commissioned by *Rethinking Construction*, asked interviewees if they could name, without being prompted, any of the initiatives for change. They were read a list of names including *Rethinking Construction, Accelerating Change*, Key Performance Indicators (KPIs) and the Egan report, and asked if they had heard of them. Unprompted recognition rates revealed that only 13 per cent had heard of the Egan report, 3 per cent Rethinking Construction and 1 per cent recognised KPIs and Accelerating Change. With a little prompting, 54 per cent of the contractors surveyed had heard of at least one of the ten initiatives listed (*Building*, 2002b).

If positive changes in the industry's processes and practices are to be realised, it is clear that the impact of initiatives such as those promoted by the Rethinking Construction initiative must make an impact at all levels within the industry. The popular perception of the industry is just as likely to be rooted in the public's experience of dealing with a small builder as they are in the image portrayed in a mega project. Thus, it is incumbent on all of those committed to process improvement to foster buy-in to such change from their supply chains and from companies with whom they have influence. Major and repeat clients have a particularly important role to play in this regard as they can use the leverage provided by their procurement activities to promote process improvement which may begin to positively impact on the public perception of the industry.

Communicating corporate and social responsibility

CSR – also referred to as corporate citizenship, is about how companies manage their businesses in such a way as to provide positive benefits for society. This can include developing responsible and ethical policies around the impact on the environment, the management of employees and on giving something back to the community in which they operate. In essence, it therefore involves businesses acknowledging the expectations of society and acting on behalf of all stakeholders. This means that businesses must act (voluntarily) in a socially ethical manner by developing policy that encompasses the core principles enshrined by CSR. Corporate and social responsibility has become more prominent in recent years as modern organisations are more vulnerable and transparent than in the past (Christensen, 2002). They are exposed to the critical gaze of pressure groups, media, business analysts and other inquisitive stakeholders. Moreover, customers want some minimum assurance that the companies behind the products or brands they are purchasing are 'behaving' properly. Einwiller and Will (2002) argue that this demand for transparency is largely supported and stimulated by the internet and purchasers who demand that companies make use of this possibility.

Despite the simplicity of the CSR concept (i.e. that caring about the community and environment in which a firm operates is likely to endear it towards the stakeholders upon which a project impacts), in many respects it remains an invention of public relations. Frankental (2001) for example, argues that it contains several paradoxes:

- *Procedures of corporate governance* – CSR philosophy is not reflected in the accountabilities of companies with regard to UK Company Law.
- *The market's view of an organisations' ethical stance* – there is no overwhelming evidence that a company's share price is affected by lack of social responsibility, even when this results in reputation damage.
- *The lack of clear definition* – corporate and social responsibility is an evolving concept and remains a vague and intangible term, which can mean dissimilar things to different people.

- *Acceptance or denial* – no corporate affairs manager will admit that his/her company is not socially responsible, yet any company that aspires to be socially responsible must be prepared to admit to its shortcomings and mistakes.
- *Lack of formal mechanisms for taking responsibility* – there are certain aspects of corporate and social responsibility where there is no compliance framework in place because there are no laws that companies are bound to comply with.
- *The placing and priority that organisations give to social responsibility* – corporate and social responsibility is often located within external affairs or community affairs departments and is seen as a function of an organisation's external relations activity. It should, however, be embedded across the organisation both horizontally and vertically.

Frankental's argument is supported by the results of a KPMG survey on corporate sustainability. Published in 2002, it revealed that of the top 100 companies in each of 19 developed countries, only 9 per cent of construction and building materials firms had submitted corporate reports on environment, social and sustainability issues. This contrasts with some 50 per cent of utilities companies and 45 per cent of chemicals and synthetics firms. Unlike other EU countries (such as France, Denmark, Holland), the United Kingdom has yet to legislate requiring firms to report on social and environmental issues. The construction industry would seem to be a long way from the social responsiveness of companies such as the Body Shop and Ben and Jerry's. As Balmer and Gray (1999) note, these companies built their core strategy around projecting a socially and environmentally responsible image.

The importance of embracing CSR as part of a corporate communications policy is underscored by the activities of protest groups which have begun to spotlight the construction industry. The World Wildlife Fund (WWF, 2003) report *Building towards Sustainability: Performance and Progress Amongst the UK's Leading House-builders* involved an assessment of 13 major domestic builders. It analysed core disclosure material (annual report and accounts) and supplementary material (company websites, formal sustainability reports, corporate and social responsibility policies, social and environmental reports and related publicity material). Despite 10 out of the 13 companies providing supplementary disclosure reports on aspects of sustainability, the report concluded that the material was 'fairly generic [with] selected examples of good practice, but tended to lack qualitative and specific performance data' (see *Contract Journal*, 2004).

Clearly, as corporate and social responsibility climbs the agenda in terms of a measure of corporate image performance, it will become increasingly important for construction firms to begin to evidence social, environmental and ethical business policies through their project level activities. Project managers must devise strategies for demonstrating a responsible attitude towards the environment, the community and company employees if they are to portray that they have a commitment towards corporate and social responsibility to external stakeholders.

Summary

The construction industry and its employers suffer from a poor corporate image. Indeed, the collective image problems of the industry are arguably manifested in the problems that the industry suffers from in attracting both its professional and blue-collar workforce (Loosemore *et al.*, 2003). This chapter has focused on how construction companies can convey a positive image to the outside world, both at a corporate level and through its projects. Portraying a positive and consistent image demands that the project manager understands the nature of the project's internal and external stakeholder groups, and devises a strategy to ensure that the organisation is seen in the best possible light by all relevant groups. From a corporate perspective, good public relations are a prerequisite for business success. The onset of corporate and social responsibility as a key indicator of business performance has necessitated the development of measured public relations strategies on behalf of construction companies. It is incumbent upon all firms to propagate a positive corporate brand, identity and image through effective public relations activities, some of which have been outlined within this chapter. However, as countless examples from the industry have demonstrated, a single project disaster can undermine years of effort in generating a positive company image. Thus, managers at the project level must be supported through appropriate training and support in media management, especially when confronted by a crisis situation.

Critical discussion questions

1 Devise a high-level corporate and social responsibility strategy for a large construction contractor operating throughout the United Kingdom. Your strategy should seek to embed social responsibility as a key driver of your firm's operational processes in order to ensure that the needs of all stakeholders are taken into account at a project level. Engaging the local communities within which projects are sited should form a key component of your strategy.

2 As a project manager responsible for a high profile hospital building in the centre of London, outline a public relations and media management strategy for the following crisis events. In each case, reflect upon how

the strategy is likely to change as the crisis event unfolds and suggest who would be involved in the damage limitation activities of the firm:

- A death on site brought about by an incorrectly constructed scaffold and faulty harness.
- Projects overspend of 50 per cent brought about by project variations linked to poor buildability of construction details.
- A project overrun of four weeks bought about poor subcontractor performance which delays the opening of the hospital and its accident and emergency unit.

Case study: corporate communication in relation to sensitive projects: the Yangtze Three Gorges Project

Introduction

This chapter has introduced several problem areas in the communicating of corporate messages to external stakeholders and has also identified the relative lack of expertise in the industry regarding such communication. It is, of course, arguable that a project manager in the construction industry is not usually employed on the basis of being expert in the area of corporate to public communication; what is required is their expertise in delivering the product successfully. However, there is a widening awareness that the industry is increasingly operating at the interface between corporations and government bodies, and 'interested' members of the public and other organisations. This interest typically arises because of the sensitive nature of the product required by corporate and governmental clients. What defines a 'sensitive' construction project is a matter for conjecture. D'Herbemont and Cesar (1998) suggest that a sensitive project can be defined in terms of high levels of human and/or technical difficulty. Technical difficulty is perhaps the most straightforward to illustrate; a nuclear power station, for example, involves some very precise and technical elements of construction in order to meet stringent performance, safety and environmental criteria. The interaction of these criteria also leads to several aspects of human difficulty. First, there is the aspect of locating and managing the different skills and expertise required for such a project, and this aspect is very much within the traditional perspective on a project manager's role. However, there is another aspect to be considered; that of the reactions of external stakeholders. In some cases, the number of stakeholders to a project may be relatively low. A small development in a sparsely populated rural location would generally be unlikely to attract the attention of anyone outside of those stakeholders who are immediately affected. Typically these will include local residents, landowners, planning officials, and others. Unless a particularly rare species of flora or fauna is threatened, or there is some significant environmental hazard involved, it is probable that the development will not attract the attention of the national or international media. Consequently, the variables in the self-images of the stakeholders may be relatively narrow and homogenous. In contrast, large and complex schemes may have a broad range of stakeholders, each of whom will receive messages in a different way. Thus, communicating to such a group en masse is particularly problematic. Nonetheless, project managers are increasingly being pushed into situations where they need to deal with this 'external' aspect of human difficulty in addition to their traditional role of dealing with technical and 'internal' aspects of human difficulty. A basic understanding of mass communication mechanisms can therefore be of value to project managers.

This case study examines the corporate communication associated with a large, complex and sensitive project which highlights the importance of many of the issues explored within the chapter. The project is the construction of the dam across the River Yangtze in the People's Republic of China (PRC), or as it is more formally known the Yangtze Three Gorges Project (YTGP). This very large construction project consistently attracted attention from the national and international media outside of PRC some aspects of the largely because it had many environmental 'threats' associated with it (hence a high level of human difficulty). This case study examines these and explores how communication was managed to its constituent stakeholders.

Background to the Yangtze Three Gorges Project

Prior to analysing the communication process in the context of the YTGP it is important to outline the size and scale of the venture. The completed project is intended to raise the level of the Yangtze river behind the dam by 175 m, producing a 'flooded' surface area of 1,084 km^2 which would inundate 24,500 hectares of farmland and displace a population of 844,100 people (both the latter figures were based on a 1992 survey so may not represent the actual final scale of the project). The dam involved the excavation of 102,830,000 m^3 of earth and rock, the placement of 27,940,000 m^3 of concrete and 463,000 tonnes of rebar, and was planned to be completed in 17 years. Along with controlling the flooding of the Yangtze, the dam is intended to produce electricity through the installation of 26 turbines with an installed capacity of 18,200 MW. The axis of the dam across the Yangtze is 2,309.47 m (CYTGPDC, 1999). All of this information has worked its way into the public domain over a number of years as the international media have focussed on data about this project. In this study, these facts have been taken from a single authoritative source so as to reduce the impact of any subjective 'noise' added during their presentation by the international media. The addition of such noise is one example of the problems that can be experienced when attempting to communicate externally about a sensitive project.

The vast scale of the project and its high profile present many corporate communication challenges for its internal stakeholders. Notably, at a national level within China, there is little doubt that the government would like the dam's image to be regarded by Chinese citizens as a positive indication of what the country can achieve. Second, there is the consideration of corporate image at the global level. In using the dam to create a positive international image, a wide range of beneficial outcomes could be achieved. However, neither of these forms of corporate image are easily achieved as mass communication, particularly with regard to sensitive projects, can be a complex business.

Potential environmental concerns relating to the YTGP

Much of the negative feeling that has arisen about the YTGP, particularly outside of China, has been focused on two key aspects of the project. The China Yangtze Three Gorges Project Development Corporation (CTGPC)[1] identified these two aspects as first, the perception of a diminished landscape and attendant cultural heritage, and second rare species degradation. One of the unfortunate aspects of this project is that it is situated in an area of considerable natural beauty that has also a long history of habitation. These factors add to the level of human difficulty that needs to be addressed. In raising the river level by 175 m a proportion of both the scenic impact and cultural heritage of the area will be lost. There are also a large number of people that will be displaced from towns and villages within the inundated area and a large amount of farmland that will be lost.

The CTGPC identified 47 rare or endangered species in the area affected by the reservoir, but claimed that because of factors such as the height above sea level at which these species existed, that no serious losses would arise from the inundation. This raises the issue of the difference in stakeholder perspective; a 'serious loss' for someone within the CTGPC may be significantly different from a member of Greenpeace, the WWF or Friends of the Earth. For example, the plight of the Chinese river dolphin

---

1 CTGPC was formally established in September 1993, with the responsibility to complete the dam and also charged with future development of the Yangtze River as an economic resource. As such, the Three Gorges dam is not the only dam planned for the Yangtze; further dams are intended upstream of the dam at XiLuodu and XiangJiaBa as part of a series of huge hydro power stations along the river.

was a particular area of debate, with the CTGPC assuring the media that the Chinese dolphin's natural habitat stopped 100 km downstream of the dam, and that the creation of two natural protection areas in the middle and downstream sections of the Yangtze would offer sufficient protection. Such actions may indeed prove to be sufficient protection, but while CTGPC are to be applauded for recognising the existence of this problem and responding to it, the organisation may have underestimated the sensitivity attributed to it by others. Environmental and wildlife groups, for example, were largely not convinced by this argument (see Talk: wildlife, 2003). A significant part of the problem appears to be that CTGPC were communicating on the basis of factors that they perceived as relevant, reliable, quantitative and valid but, which the receiving groups appear to have perceived as irrelevant, unreliable and largely invalid.

Communicating the YTGP benefits

Mitigating the main negative impacts acknowledged by the CTGPC required a particularly effective corporate communications approach. The aspects of the corporate communications process are considered below under headings derived from Figure 3.3, which signifies the content and context of the communication process. The analysis focuses on an information booklet developed by the CTGPC as an authoritative basis for presenting the project to the external environment, the contents of which are discussed here.

The sender

Considering that all explicit attempts at communication should start with the sender (or communicator), the analysis of the communication process with regard to the YTGP should commence with the specific consideration of their role. In this case, the communicator will be taken as the author(s) of the CTGPC's text in relation to the project, although this text is not attributed to an individual. The analysis therefore has to regard the corporation as the author, which then raises questions concerning the author's objectives in producing the text and its perspective on the project based on accepted values and beliefs. A more detailed analysis suggests that the message that is intended to be communicated will almost certainly be the result of significant pressure to conform to a particular objective. This is that the dam will be a huge economic success and, flowing from this, will have a positive effect on factors such as rural poverty and quality of life, water quality due to reduced pollution, significantly lower emissions of carbon dioxide and other greenhouse gases (as hydro-electric power displaces primitive coal-fired generating capacity), and improved irrigation for farmland around the inundated area.

The message

Having examined the sender, the next stage is to consider the message itself, with particular regard to the content. There are two aspects of the message to consider. First, the selection and structuring of the message content and, second any pressure or constraint exerted by the message. Both of these considerations are effectively referring to the process of gate-keeping, which is commonly practised by organisations and individuals employed by them. Gate-keeping is essentially about the selection or rejection of content for a message and is based on control of the communication channel and medium. The value decision regarding the exerting of such control as being either 'good' or 'bad' will not be entered into here as it is largely a matter of self-image and thus, for the individual receivers and senders to decide upon. The important aspect of gate-keeping is that decisions are made as to what information (content) a particular message to a particular receiver (individuals or groups) will contain or, perhaps more significantly, will not contain. The CTGPC booklet for example contains quite detailed quantitative data, but omits information on the potential for social unrest and resistance to the development of the Yangtze by the construction of a dam. Unless such information can be communicated by another channel, perhaps one outside the control of the initial communicator, then it is possible that the gate-keeping process will ensure only positive content is contained in messages to receivers. With the development of mobile phone and internet connections technology, it is becoming increasingly difficult for any organisation to practice absolute gate-keeping. It is perhaps therefore more appropriate for such organisations to develop skills in responding to spontaneous feedback from the receiver(s). In order to do this effectively, there is a need to analyse the medium used to convey the message to the receiver(s).

The medium

Consideration of the medium for communication results in an awareness of the suitability each type has for a particular type of message. Each medium has its own characteristics, some of which may be shared with other media while some will be unique to that method. The medium of radio, for example, is fine for communicating the spoken word and music (any message that is aural in nature) but is of no use in communicating a message that is largely or wholly visual in nature. Radio broadcasters sending reports when on location always have the problem of trying to set the scene about what is going on around them so that their message can be put into some sort of local context by the receiver(s). Television broadcasters can simply point at images around them on screen to set the local context and thus, can communicate detail more rapidly. In the case of the CTGPC booklet, the medium is the printed form; a 29 page booklet of explanative text, detailed tables of data and several figures of construction details etc. For a receiver with a reasonable level of experience of the construction industry, the document is informative and generally communicates the details of the project clearly. This particular style may well be based on some research regarding the image of any receiver willing to purchase such a booklet. Anyone wanting a cultural history of the Yangtze region would not buy a document that deals with engineering and management issues relevant to the dam's construction. This introduces the factor of differentiation within the communication process. The booklet by the CTGPC has a distinct engineering and management flavour, which will encourage a particular type of receiver, but may discourage other types.

The receiver

On this side of the communication model there is a need to reconsider aspects of the medium, the nature of the receiver and any opportunities for spontaneous feedback. The issue of feedback is a relatively simple issue to deal with in the case of the formal CTGPC document; there is no opportunity for any spontaneous feedback. In some respects this is a pity as feedback is of value to the sender, if only to help in establishing if the message is being received and if so, whether the reception is a favourable one. In the event that the feedback is not favourable, the sender has the opportunity to modify the message as it is being communicated. However based on the assumption that the document was not the result of a consultation process, in this case the only opportunities for modifying the message are through rewriting any new editions of the document or in releasing a further document with the specific intention of addressing the areas of the original message that were not received favourably. Furthermore, the closest that can be achieved to spontaneous feedback in this case would be if the receiver was able to communicate directly with the sender. Whereas it is theoretically possible to immediately draft an email, there is a need to establish an address to send it to. Similarly, with a telephone or fax message there is a need to establish the sender's telephone number. In the case of the CTGPC document, neither of these pieces of information is supplied. Thus, the sender is missing two opportunities for rapid, if not truly spontaneous, feedback. Hence, this type of mass media-based communication is constrained in terms of its ability to elicit a meaningful response from its intended audience.

Given the emphasis placed on the importance of the medium (see section on 'The medium'), it is worth exploring its relationship with the receiver. There are several possible responses to the medium and the message. First, there may be a process of selection in which the receiver initiates what is referred to as 'selective attention'. The general consensus of research in this area is that people pay attention to those messages that are most in tune with their own self-image and beliefs. This is the first aspect of selectivity that the communicator must either overcome or make use of in communicating their message. The second aspect of selectivity is that of selective interpretation, in which the receiver starts to analyse the selected components of a message. Selective interpretation is carried out on the basis of the receiver's values and beliefs, education, communication skills etc. All of these factors bring about an interpretation of the message, and it is possible that it will not be the sender's intended interpretation. In the example of the possible threat to the River Dolphin, for example, it is quite possible to construct an interpretation of the stated population figures that is based on possible inaccuracies in the methodology of data collection. After all, just as species become extinct, examples of species that were thought to be extinct have been encountered. Within a mass media (or corporate communication) context, the identification of each individual receiver's standing with regard to such factors is highly problematic. One strategy for dealing with this is to send a highly constrained message using specific language or images that is targeted at a well-defined and understood group. The CTGPC document has attempted this to some extent in that its use of language and data content indicate it is not intended for general consumption. The document is arguably targeted at the more technically minded sections of society who also have a clear understanding of the majority of the terminology used.

Finally, there is a third aspect of selectivity for the message to respond to; selective retention. This infers that people choose to remember certain experiences and try to forget others. This cognitive dissonance occurs when an individual encounters a message that is not harmonious with their beliefs or values. They may accept that there is sufficient evidence in the message to convince them that their own beliefs and values are now wrong, or can choose to continue to assert that their values and beliefs are correct. Only after these three aspects of selectivity have been played out can the receiver begin to consider the message in terms of their own unique interpretation of it. This interpretation is one of the effects of the medium and message. Other effects, such as immediately deciding to initiate a campaign of support for, or agitation against, a project such as the YTGP, are possible.

The context

Having dealt with issues of receiver, there is a need to consider one further aspect of the medium; the context in which the receiver experiences it (see Figure 3.3). As in the consideration of the medium by the sender, there is also a need to consider *how* the receiver will experience the selected medium. A medium such as television, for example, provides an immediate experience that cannot be returned to (unless the message is in some way recorded). A book, on the other hand, allows for an experience which is more within the control of the receiver. Books can be read or ignored, but if they are ignored initially, there is always the possibility of reading them at a later date. Once the decision is made to read the book, the process can be approached in a variety of ways. One possibility is to read it from cover to cover in a linear manner and at one sitting. For the less committed, it is possible to read cover to cover in several sittings, with each reading taking place when the reader chooses to engage with the book again. Alternatively, the book can be scan-read, with only the most interesting sections being read in detail (see the discourse around selectivity earlier). Consequently, the sender can be seen to have little control over the manner in which the receiver experiences such forms of communication and so the medium does have to be suitable for the intention(s) of the message.

The social environment in which a message is received is also a significant contextual variable. It is very rare that messages are received in isolation of other data, information or stimuli (noise), all of which will add to an individual's interpretation in the processing of the message. An individual's values and beliefs form part of the consideration of the social environment, but it also comprises other factors. The receiver's age group is important, for example, in that this affects what people see as important – whereas those under 30 may not generally be concerned about pensions for example, those over 50 might be. Similarly, the intended receiver group for the Yangtze Dam booklet may not be generally concerned about environmental issues and species protection, but may be concerned with opportunities for wealth generation and increased economic activity. Other factors typically included in this area are those of family background (values and beliefs may be consistently passed between generations), ethnicity (and culture) and religion.

Conclusions and lessons learned

It is important to recognise that large projects occur throughout the world and that they are not always well-received by all stakeholders, Twyford Down is a UK example of a large project that raised considerable concerns about damage to a natural environment and species loss. Some unpleasant scenes of conflict between contractors and protest groups were given considerable coverage in the media both nationally and internationally, neither of which will have helped create a positive corporate image of the companies involved in the eyes of many external stakeholders. The management of communication between the project members and all external stakeholders in such cases is a key contributor to the operation of the project and cannot therefore be approached on the naïve belief that nobody is going to pay much attention to a few protesters. The alternative however, requires the input of considerable energy to underpin effective communication.

This case study has revealed a particular approach to mass communication undertaken by the corporation with responsibility for the project (the CTGPC). When considering the success of the approach adopted, it is appropriate to consider the first image, fact or concept that comes to mind at the mention of the Project. It may be that, prior to reading this case study, the reader had no knowledge of the project or what it intended to achieve. There is certainly a low level of probability that anyone reading this book would have a concept of the project as a representative 'brand' for China. However, considering that the Three Gorges Project is simply the first of at least three dam projects on the Yangtze, there would seem to be a significant opportunity to create a brand as part of the mass communication process. Developing such brand recognition would require those responsible to adopt the principles of mass communication discussed in this chapter. Creating a positive brand, identity and image through a project (or set of related projects) requires a considerable effort combining a number of different media, all of which should portray the same positive messages but in a way which accords with the particular perspectives of the individual stakeholder groups involved. In this way, the technological ingenuity, positive environmental benefits and size and scale of the project can be used to offset the concerns of those with concerns as to the appropriateness of the venture.

Part III

Future directions for construction communication

This part briefly explores the future of communication in the construction industry. It examines the role and importance of information and communications technologies as enablers of communication within the industry both now and in the future. There are also suggestions as to how the industry might begin to address its previous failings by drawing upon the principles expounded within the book. The conclusions provide the reader with suggestions as to the future communications challenges for the industry and the directions that research could take to try and overcome these issues.

Chapter 8

Information and communications technology

Introduction

In the 1950s and 1960s, research emphasised face-to-face interaction as the primary method of communication in business and organisational life, but today the nature of communication has been irreversibly changed by the onset of communication technologies (Eisenberg and Goodall, 1993: 204). When ICT is considered in the context of construction, the probability is that those not connected to the industry will perceive there being little association between leading edge technology and an industry which is still associated with traditional processes. However, the use of ICT now shapes many of the industry's design and production functions, as well as having a significant impact on the operation of finished buildings. For example, one area where ICT now dominates is in design, where computer aided design (CAD) dominates over traditional manual drawing methods. Similarly, project-planning software is routinely used on all but the smallest of projects and allows for the careful scheduling and monitoring of the production function. In the operation of modern built assets, energy management systems often provide automated methods for controlling buildings, and computers now facilitate the scheduling of maintenance and facilities management activities. Thus, it is increasingly being realised that the effective use of ICT is necessary for delivering efficiency and improved project performance in the construction industry.

Rather than focus on individual applications or specific technologies (for which existing textbooks abound), this chapter explores the current and potential role for ICT in facilitating communication within the industry. The opportunities and constraints on the uptake of technology enabled communication are explored in order to establish how the industry could better exploit its implementation in the future. Furthermore, the implications of ICT-enabled working are examined, particularly in terms of how the industry's people and processes might be affected in the future.

The utilisation of ICT in the construction industry

In today's business world, it is widely acknowledged that information systems are essential for organisations to survive and prosper, as they provide the capacity for firms to offer new products and services, operate in remote locations and develop the ways in which they do business (Laudon and Laudon, 2002: 4). Information technology, with its associated fields of telecommunications and microelectronics, has revolutionised the ways in which information is managed and people communicate within industry. According to Fryer *et al.* (2004: 82) it has achieved this in construction in three principal ways:

- by speeding up the processing of information;
- by making information more easily accessible;
- by improving management information systems for more effective decision-making and control.

A management information system (MIS) is an organisational and management solution, based on information technology, which addresses the special challenges posed by a particular environment (Laudon and Laudon, 2002: 11). Integral to the effectiveness of an MIS is the communications technology, consisting of both physical devices and software, which allows the transfer of data from one location to another. By using ICT in an integrated manner, such systems can facilitate information flow, allowing users to transfer greater volumes of information rapidly and efficiently. New ways of working based around developments in MIS have placed greater emphasis on the way in which people exploit the increasing opportunities presented by ICT. However, despite the potential benefit of ICT, convincing construction organisations to embrace its use and implementation has proven a difficult task in comparison to the automotive and aerospace industries (Duyshart *et al.*, 2003). Indeed, the UK construction industry is periodically criticised for being too slow to adopt both new technologies and methods of working. Reasons as to why this should be the case are almost certainly rooted in attitudinal resistance to technology, physical constraints provided by project-based working and the prevalence of SMEs within the industry, many of which do not have dedicated ICT specialists. These have provided problematic structural and cultural conditions for an ICT revolution to take place within.

For several years, many within the industry have believed that multimedia communications are set to revolutionise construction, with virtual reality used at scheme design, site meetings conducted through an intranet and drawings transferred back and forth via email (Bunn, 1997). However, from a project manager's perspective it is important that both the potential of ICT to improve communication processes *and* the reasons as to why its application often fails in construction is explored. This knowledge will allow them to use ICT in a way that is aligned with the industry's cultural predisposition towards the adoption of such technology, rather than attempt to foist inappropriate technologies on supply chains that are unable to benefit from approaches.

Virtual working as a driver for ICT innovation

A key driver for the use of ICT in communication has been the onset of what has become known as *virtual teams*. The nature of teams within the construction industry has changed significantly in recent years, primarily because of changes in organisations and the nature of the work they do. Many large organisations have become geographically and sectorally distributed, thereby working across different industries and transcending national and international boundaries. This, coupled with the fact that customers, suppliers, and other stakeholders are now more involved during different phases of the project life cycle have meant that conventional teams are no longer able to satisfy the needs of multinational organisations. This has resulted in the emergence of virtual teams, which can be defined as:

> a group of people who interact through interdependent tasks guided by common purpose... Unlike conventional teams, a virtual team works across space, time and organisational boundaries with links strengthened by webs of communication technologies.
>
> (Lipnack and Stamps, 1997)

Virtual teams create the perception that the team members work together in the same space and time with the same set of organisational norms (Cantu, 1998). However, team members do not need to work in the same organisation, or be co-located in the same place.

According to Raghuram *et al.* (2001), three main factors have triggered the emergence of virtual working. First, there has been an important change in clients' requirements. Within construction, for example, they now demand a full range of services: design, build, repair and manage. This, in turn, requires a change in thinking on the part of designers, contractors and suppliers, who now have to work together to fulfil client needs. This change includes a cultural/attitudinal shift that moves from product-driven through a service-driven to a knowledge/technology-driven strategy to fulfil developing client needs (Johnson and Clayton, 1998). Second, information technology is now ubiquitous within developed societies. Inexpensive computing power, extensive networks and the Internet have provided employees access to the information they need to perform their work in locations other than traditional physical office spaces. A third factor concerns globalisation of the marketplace. A trend towards corporate restructuring can, in part, be attributed to an increase in corporate layoffs, mergers and acquisitions, competition and globalisation. Work groups with a dispersed membership (virtual teams) have therefore become critical for companies to survive (Lurey and Raisinghani, 2001). Increasingly, organisations are trying to grow in size to carry out large projects in different parts of the world using dispersed membership groups. Globalisation of the marketplace makes virtual teams the primary operating units required to achieve a competitive advantage in an ever-changing business environment (Lurey and Raisinghani, 2001).

An important feature of virtual working is the heavy reliance on ICT. The new forms of ICT provide the foundation for the effective operation of virtual teams that would have not been feasible even a decade ago (Townsend *et al.*, 1998). ICT innovations have enabled the development of what have become known as *virtual organisations*. Whereas virtual teams are seen as a temporary arrangement (Jarvenpaa and Leidner, 1998), or a project/task focused group that dissolves after the project is completed (Townsend *et al.*, 1998: 17), the virtual organisation is a long-term arrangement with long-term common interest/goal (Ahuja and Carley, 1998). The development of such enterprises is arguably fuelling an ICT revolution, the effects of which are being felt within the construction industry.

The evolution of ICT in construction

The cost of computing technology has fallen sharply in recent years, with even the smallest organisations now being able to adopt powerful ICT that would have previously been beyond their budget in the past (Baines, 1998). This has enabled ICT to become commonplace throughout manufacturing, service industries and government. However, with ever-widening possibilities comes the problem of providing meaningful definitions of what computers are, how they work, and perhaps most importantly, how they are used. What was once simply known as information technology, has now evolved into the rather grander sounding *information and communication technology* (ICT). This infers that information technology is now effectively acting as an *enabler* of communication within modern organisations and society more generally. Such technology is arguably revolutionising the ways in which people communicate with each other and the ways in which they manage information flow. It has even re-shaped the processes with which people are involved and the data that they use to undertake their functional roles. Consider the following examples of where technological developments over the past few years have fundamentally changed the ways in which construction site operations are conducted:

- *Facsimile machines* – the advent of this communications device allows for the rapid transfer of information, instruction and orders between remote sites and other offices. Before facsimile machines were developed, most communication had to travel via post or courier. It is interesting to note that contractors frequently use this medium to request information from architects, and ask for a reply to be faxed back.

- *Mobile telephones* – these have enabled users to communicate from just about any site location, even when they are engaged in physical work activities. The advent of cheap, portable communication technologies has arguably fundamentally changed the way in which small businesses operate in particular, as it has removed the need for an office-based presence. The associated functions such as text messaging, photographic functions and even video capture also have massive potential for impacting upon the way in which construction firms operate. *Contract Journal* (2004c) reported on a survey undertaken by the Mobile Data Association that revealed that one in six (17 per cent) construction industry managers said they actively use text messaging for business communication. Apparently, telephone numbers and reminders of instructions feature regularly as messages, with surveyors and property consultants making use of picture messaging.
- *Email* – this has allowed for the rapid transfer of both textual and graphical information to speed up decision-making and problem-solving in the construction process, as well as providing a convenient way of rapidly transferring written orders and instructions. Wireless networking technology has overcome many of the problems previously associated with utilising email in site-based locations.
- *The internet* – E-commerce has enabled orders for materials to be placed using web-based systems. The Internet has also allowed for central repositories and databases of information to be developed which can be accessed by any project participant (see the case study at the end of this chapter). The potential of the Internet is, perhaps, only just being realised in construction, but the potential for E-business is undoubtedly significant for the future operation of the sector.
- *3D CAD* – 3D modelling has facilitated the checking of production information in order to mitigate risk from construction operations. For example, it can be used for clash detection in highlighting where service installations would be impeded by the structure of the building and vice versa. Virtual reality applications take this a step further in allowing building designs to be modelled in any number of different dimensions. For example, the addition of time as an additional dimension to the spatial data enables users to review the progress of work against project plans and to identify elements of the works where time overruns are likely to occur.

The above examples show increasing levels of technological sophistication, but all are now relatively commonplace within the construction industry. Clearly these tools are not necessarily technologies which have been developed to account for the specific needs of the construction industry, but represent the appropriate application of mainstream technologies to the unique context and challenges that the sector presents. It is important to note, however, that the implications of the change that such technologies are bound to induce is understood if their potential is to be harnessed for the collective good of the industry. Simply adopting a new information technology without considering the wider implications for an organisation and its operating context can have damaging consequences as will be explored later in this chapter.

In broad terms, communication technologies can be classified into two categories (Huber, 1990):

- *Computer-assisted communication technologies* – image transmission devices (e.g. facsimile, video conferencing etc.), electronic and voice mail etc. In essence, these are designed to speed up communication and information flow, thereby bringing people together more rapidly than having to travel to physically meet.
- *Computer-assisted decision-aiding technologies* – online MISs, inter and intranets. These technologies are designed to provide access to information required for making decisions.

Both forms of communication technology can be seen to have benefited the construction industry in recent years. Computer assisted communication is now used by virtually all construction firms in some way (e.g. through facsimile and email) and most now have access to online information sources to aid their decision-making processes.

Some examples of ICT innovations in construction

To review all of the ICT applications currently utilised in the construction industry would present a massive task and is well beyond the scope of this book. Nevertheless, it is worth briefly reviewing some of the major ICT innovations that have fundamentally affected the ways in which the industry's processes are managed and more particularly, the ways in which people communicate within the sector. Examples of both computer-assisted communication technologies and computer-assisted decision-aiding technologies are briefly discussed here.

Computer aided design (CAD) and virtual reality (VR) applications

Contemporary CAD packages range from freehand sketching tools through to full-blown 3D systems capable of carrying out analysis of everything from the quantity of concrete through to the predicted behaviour of structural elements under load. The earliest CAD systems were used by the engineering industries, particularly in the automotive sector, but it was not until about ten years after the first systems became available that equipment costs had decreased to the extent that architecture practices could consider them financially viable. However, by this point the level of usability had improved in that CAD had moved out of purely numeric operations (cost estimation and structural analysis) and into more mainstream design operations. By 1989, a RIBA survey showed that 25 per cent of all practices used some form of CAD, 66 per cent of practices employing over 11 people made use of 2D CAD, and 50 per cent of the same group utilised 3D versions (RIBA, 1990). By 1993, over 90 per cent of this group were using CAD for predominantly 2D work (CICA, 1993), and the rising trend has continued. Increasingly, CAD packages are being supplemented by visualisation or VR software which can allow buildings to be modelled in many more dimensions. Using such packages to convey production information has massive advantages over manual techniques and made the entire project delivery process more dynamic and responsive to change. The latest technology allows performance data to be dynamically linked with production models. This enables the impact of design decisions and production information to be assessed for its time and cost implications, as well as a range of other 'softer' performance criteria.

Project planning and estimating tools

A second well-known area of ICT usage is in project planning and control packages. Again, these range in sophistication from producing the ubiquitous bar (or Gantt) chart (in a wide range of presentation formats), through more sophisticated network and critical path analysis packages to advanced software which allow 'what if?' analyses as an aid to project planning (see Retik and Langford, 2001) and risk management (see Dikmen et al., 2004). These popular applications have now been supplemented with a vast array of supporting applications which can be used to facilitate the design and construction process and the interfaces between them (see Aouad and Kawooya, 1996). For example, estimating packages can also be placed in this group, in that they are essentially a control tool and in most cases have planning functionality allowing the modelling of cash flow against the construction programme. From a communications perspective, such tools facilitate the assimilation of complex project information and support the integration of complimentary datasets. This, in turn, allows more sophisticated project monitoring, such as cost/value reconciliation exercises, using real-time project information.

Further developments involve the modelling of the construction process at design stage and the employment of a logistics management specialist to assist in pre-contract planning and construction phase control. Typically, this involves taking a 3D drafting package and incorporating a construction – programming tool. The virtual model can be continually updated during the construction phase with software linking the materials requirements to Architectural 3D construction planning software (Building, 2001d).

Communication of project productivity and performance

In recent years, there has been a proliferation of sophisticated tools for measuring and conveying project performance information. Such tools are able to manage project performance against a range of different dimensions which together provide a more robust way of ensuring the successful outcome of projects. A good example of an integrated suite of tools for achieving this was developed by the Building Research Establishment's Centre for Performance Improvement (CPIC). Their toolkit (entitled CALIBRE) comprises five main parts (Time Evaluation and Assessment Measurement System (TEAMS); Building Research Establishment (BRE) Plan; Site Environmental Assessment System (SEAS); SMARTWASTE; and AS-built (CD Rom). The toolkit uses various modes of communication to collect, analyse and present useful information to project teams, including clients, and has been successfully employed by Sainsbury's, Railtrack, Whitbread, BAA and McDonalds restaurants (see CALIBRE, 2000). TEAMS, for example, identifies how time is used on construction sites and separates time into added value time, support time, statutory time and non-added value time. It determines cause of wastage and allows the project team to develop an action plan for eliminating non-added value time. Carr and Winch (1999) reported on the use of this activity sampling technique on two projects, one in the United Kingdom and one in France. They provide an account of the four elements in applying TEAMS – mapping the construction process; identifying and coding; monitoring the site; and analysis, reporting and feedback. Essentially, the process involves a trained observer recording the activity of work trade contractors on site and inserting data into a hand-held data logger. The application of TEAMS assists in making the contractors master programme a 'live' document that could be updated with real-time evidence of progress. Considering that trade package managers normally collect progress evidence by a combination of looking (at the site), reading (as built labour and material schedules) and listening (to trade package gangers/foreman) the use of TEAMS may provide a more robust assessment of individual trade package progress. In addition it is intended to act as a process improvement tool and can be used to provide evidence of where corrections can be made. Thus, measurement tools such as CALIBRE offer powerful tools for ensuring the achievement of project objectives through the real-time management of projects.

Internet-based communication resources

The Internet (or Worldwide Web) has opened up a new range of business possibilities which were hitherto impossible to deliver within a production oriented industry like construction. Web-based applications can connect organisations through a common medium, which provides an opportunity to integrate communication flow within the project supply chain. Internet-based interactive business tools are now available which have the potential to significantly improve the ways in which project participants communicate (Duyshart *et al.*, 2003). Although construction is yet to embrace 'E-business' to the extent of other service oriented sectors (see Emmitt and Gorse, 2003: 96), there are signs that several forms of Internet-based business are emerging as popular ways to working in the sector. Examples include, E-procurement (to facilitate the purchase of products and components), E-commerce (supporting transactions between the vendor and purchaser) and E-collaboration (enabling communication amongst supply chain partners). As a case in point, the industry has been fairly enthusiastic in its adoption of information and communication technologies to improve collaborative working; recent research shows that the industry is currently using over 20 Internet portals to share project information online (France, 2002). The real advantage that such systems offer is speed of access to information and products and a common frame of reference for doing business. It is therefore likely that computer-assisted decision-aiding technologies are likely to take more of a foothold in the future within the sector.

Although examples of where Internet-based project management systems have yielded tangible benefits to project performance are currently relatively few and far between, recent research has offered some insights into the types of approaches that may offer potential for the future. Thorpe and Mead (2001) investigated the adoption of Project Specific Web Sites (PSWSs) as a means of overcoming the sequential flow of information to and from each member of a project team. Their review of three construction projects found that where PSWSs were utilised, project participants pulled information from a single central source. This allowed project participants to circumvent traditional chains of command, eliminating many of the communication barriers (duplication, misunderstanding due to sketchy teamwork). However, despite the ability to speed up the flow of information, PSWSs can cause information overload. Thorpe and Mead suggested that key project participants (architects, project managers, site superintendents and site office engineers) must all utilise this technology if employed on a project because when one of these participants refuses to participate, the system will then lose its effectiveness. In another study, Weippert *et al.* (2003) investigated four case study projects in Australia where *Internet-based Construction Project Management* (ICPM) systems had been used. It was found that although use of ICPM solutions were generally perceived as convenient, inexpensive and fast, it was not possible to determine conclusively whether they positively influenced the nature of communications between the project participants. It was concluded that more research is needed to explore how to overcome industry *cultural* barriers traditional work habits that may prevent the widespread adoption of innovative ICT tools and ICPM communication systems. Thus, it would seem as if attitudes and ingrained ways of working may still offer a significant barrier to the uptake of web-based technology within the industry. This will be explored further later in this chapter.

Supply-chain management tools

In recent years the construction industry has adopted Electronic Data Interchange (EDI) technology that is normally associated with other industry sectors and has applied it in association within a logistics management philosophy. Several clients and contractors have piloted innovations which combine wireless technology, electronic tagging, hand-held devices, web applications and global positioning systems (GPS), which together allow materials and components to be tracked during construction. Benefits of such technology include a reduction of lost delivery notes and payment delays, improved material management and a reduction in defects. Encoding hypertext URL addresses within electronic tags allows web-enabled hand-held computers to access manufacturer's data and so has huge potential for future E-commerce transactions (see Constructing Excellence, 2004a). Table 8.1 summarises the benefits that can be derived from the employment of ICT in relation to supply chain improvements.

Wearable computers

One evolving area of ICT of particular significance to construction is the development of 'wearable' computers. In other sectors, such technology has begun to impact on the way in which information flow is managed to the extent that massive efficiency gains have been achieved. An illustration of how wearable computers can impact positively on work productivity can be found in one of the products manufactured by Xybernaut (2004), a smart display by the name of Atigo. This product was used to deal with a problem faced by a company transporting a high volume of goods through the US rail system. Prior to the use of Atigo, the company's system involved printing out work sheets at the start of every shift and hand-written delivery notes (transcribed at a later point in the process). Using this manual system, the recoding of the data typically lagged 4–8 hours behind actual deliveries. The system resulted in high costs related to data processing and increased risk of transcription errors, as well as being unable to respond to customer queries in real time. Atigo is a wireless display unit that offers full roaming functionality within the train conductor's normal working environment. It can be comfortably worn by the user and operates via a sensitive touch screen. Since adopting Atigo, the company has found that it is now possible to produce real-time updates of the progress of deliveries, there are fewer operator errors, increased productivity brought about by benefits of eliminating time and effort in paper-tracking, and faster access to customer data. Evidence of wearable computers in the construction industry is not easily uncovered, but it is easy to envisage the potential of such devices within a construction context. For example, it is possible that site hard-hats will eventually incorporate 'head-up displays' that will display plans and drawings pointing out exactly where the workers need to be and what needs to be done (see *Construction News*, 2000).

Potential barriers to the uptake of ICT in construction

Although construction can be seen to have benefited from the development of communication technologies discussed above, there is little evidence of the widespread application of leading-edge systems to facilitate communication within the sector. Indeed, there is more evidence to the contrary. It is important to speculate, therefore, why such problems may exist. There are many practical problems inherent in sharing information between the various disciplines involved in a construction project. For example, architects and engineers often favour CAD systems, whilst building services engineers tend to use more specialist software packages which makes information sharing more difficult. The implementation of computer-based management systems can also create problems for an organisation, such as negative attitudes from employees to new ICT-enabled processes, disruption to organisational processes or even the undermining of established informal networks and communications (Fryer *et al.*, 2004: 83).

In exploring what the barriers to the greater take-up of ICT are within the sector, it is useful to draw upon research studies which have identified constraints on particular types of communication technology within the industry. Ganah *et al.* (2000) investigated the use of computer communication and visualisation during the construction stage of medium to high-rise buildings. This was primarily in respect of the use of such technologies within contractors and consultant organisations when communicating with other participants in the design and construction process. Their research revealed that the use of computer visualisation and communication was very low, with the most common methods of communication between design and site teams utilising traditional methods and tools such as 2D drawings, face-to-face meetings, written statements, telephone and fax. Ganah *et al.* (ibid.) argue that these traditional approaches are insufficient

Table 8.1 How RFID could change the face of construction

Areas of improvement in supply chain	Wireless technology	Tags/barcodes	Handheld devices	Web application	GPS	Benefits of technology
Tracking manufacturing process in real time	*	*	*	*		More efficient, effective and profitable process
Producing mass bespoke components in the timescale and cost of mass produced components	*	*	*	*		More efficient, effective and profitable process Allow expansion of product range
Tracking delivery of goods	*	*	*	*	*	Improve traceability of goods
Paperless invoicing system	*	*	*	*		and reduce number of disputes on whether goods have been delivered
Internet visible ordering and delivery service	*	*	*	*		Eliminate the use of paper-based system Improve customer satisfaction
Asset management	*	*	*	*	*	More efficient effective and profitable process
Locating material in the factory and on site	*	*	*	*	*	Correct goods located and will save man hours in locating goods

Providing up-to-date information on site for a tagged building product, for example on health and safety, installation, etc. Enable dispute resolution Demolition/deconstruction	*	*	*	*	This will be possible when the bandwidth Increases for hand-held Internet-enabled devices. Provide a means to deliver drawings and information required in real time and assist identification and installation of correct parts. Improve information flow in the entire supply chain and aid problem- and dispute-resolving
Maintenance of building services and assets	*	*		*	
Facility/asset management	*	*	*	*	Reduce maintenance inspection times With increased bandwidth wireless technology to be used to download data required by operatives in real time and improve efficiency, accuracy and reduce number of visits
Aid in integrating supply chain	*	*	*	*	All the above

Source: *Construction News* (2004c).

for dealing with buildability problems, but concluded that the professional firms surveyed were reluctant to adopt visualisation techniques because of custom. These research findings suggest that attitudinal and cultural barriers may impede the take-up of ICT solutions more than technological constraints. Given the plethora of evidence which suggests that small and medium-sized construction businesses are ill-prepared for the increased application of ICT (Love et al., 2000; Ng et al., 2001; Wong and Sloan, 2004), it is likely that resistance will be found throughout the industry.

If barriers to the take-up of ICT are apparent at an organisational level, it is reasonable to assume that resistance will be even greater at a site-based operations level. This contention was explored by Bowden and Thorpe (2002) on the M6 Toll motorway project. Seventeen construction worker volunteers were given various hand-held computer devices to complete four usability tasks so as to find out if different devices would suit different people and tasks. As well as providing feedback on the ease of completing the four tasks, the volunteers were also required to assess the portability, screen clarity, appearance, and ease of data entry and input keys on each device. This study identified several barriers that may discourage use of this technology, but contrary to common perception, site-based staff were ready and willing to use devices to communicate with the wider team. Thus, it may be the case that resistance to ICT is rooted at higher levels than might be expected. The successful uptake of ICT therefore requires appropriate strategic decision-making by top management if it is to be successful (Whyte et al., 2002).

The examples discussed earlier suggest that the wholesale implementation of ICT within the industry is likely to be fraught with difficulty. The change that ICT induces is likely to generate increasing resistance, particularly as the capabilities and possibilities of ICT-enabled processes grow and evolve. Thus, despite the undoubted advantages of ICT, barriers to its uptake are beginning to create a so-called 'digital divide' between those who are digitally 'enriched' and those who are digitally 'impoverished'. Whilst some construction firms have embraced ICT and the advantages that this brings to managing processes and projects, others have remained wedded to traditional approaches to communication, content that tried and tested approaches are adequate for meeting their particular needs. This phenomenon is not peculiar to construction, but cuts across ICT usage in society, as is evidenced by the ongoing debate in the United Kingdom over broadband provision. It is important to ensure sufficient bandwidth to enable users to access and use online learning. However, only a relatively small number of Internet users currently have broadband connections in their homes despite the advantages that they offer. The same principle applies in an industry context, in that firms operating using limited ICT systems will find it increasingly difficult to interface and work alongside those with those who have fully embraced the latest technologies.

Explaining the industry's reluctance to embrace ICT: differential adoption

Earlier in this chapter the development of CAD and its adoption by the UK construction industry was presented. This was used to illustrate the manner in which the industry has adopted one form of technology-enabled communication. However, in the context of a single form of ICT (CAD), this illustration may present a model of adoption that could be assumed to be consistently replicated in all sectors of the industry and with all forms of ICT. Research by Moore and Abadi (2005) suggests that this is not actually the case as certain forms of ICT are consistently adopted, while others are adopted only by certain sectors of the industry. It appears to be the case that some of these sectors represent a level of usage so low that no viable claim can be made with regard to the industry having comprehensively recognised any benefit for the ICT form in question. This introduces various factors with regards to the implementation of ICT, especially in terms of the rate at which construction firms adopt ICT applications.

A good example of the differential manner in which ICT is embraced by the industry can be found with regard to the adoption of virtual teams (see earlier). The industry has failed to comprehensively adopt this particular form of working, largely because they believe that it is nothing new and is simply a repackaging of the manner in which they have always worked (Moore and Abadi, 2005). Whereas it is certainly true that the industry has a history of distributed team working, they have been constrained in their effectiveness by issues of minimal horizontal and vertical communication that could be facilitated by ICT applications. Thus, it would seem that the concept and operation of virtual teams is an ideal concept for the industry to embrace, and yet their development remains stymied by its apparent inability to adopt appropriate technologies and ways of working.

What emerges from this discourse is a picture of the industry as one where different types of organisations adopt technologies at differential rates. The corollary of this is likely to be difficulties in collaborative working amongst firms connected by being members of a common supply chain. Possible reasons for this include:

- *Industry fragmentation* – a possible factor in explaining the differential picture of adoption of ICT across the industry is its fragmented nature. This is characterised by a high level of specialist organisations, each with their own functional and professional objectives. This inevitably results in differing levels of adoption of new technologies and moreover, to an incompatibility of the systems used.
- *Limitations in bandwidth* – The nature of information flow associated with construction processes is one that is rich and hence demands significant computing power to transfer the information required.

This in turn demands that all users have access to appropriately powerful computers and moreover, broadband connections to allow the rapid transfer of information. Given the plethora of small companies with limited ICT capabilities, it is likely that few construction supply chains will be sufficiently sophisticated to allow for the rapid and seamless transfer of large data files.

- *Cost* – Investment in new ICTs is expensive for an industry which has historically operated on relatively low profit margins. The cost of the technologies involved in supporting the provision of increased bandwidth, for example, requires the organisation to examine what forms of corporate information it deals with. This will typically include documents, messages, conversations and more structured data. It is increasingly expected that organisations will not only store all of this information, but also that they will provide their people with the infrastructure to immediately access all of it. The advent of electronic infrastructures has initiated an explosion in the volume of information accessible and hence, the cost of enabling technology to support it. However, the key cost may not be the outright purchase price of the technology, but rather the indirect cost of learning what the technology will do and how to utilise its functionality effectively.

- *Information overload* – The 'Law of Data' states that data will expand to fill the space available for storage; as more memory becomes available on a computer system, data will expand to fill it. Thus, it is highly likely that as rich information can be transferred more rapidly and efficiently, it will eventually reach a stage whereby those expected to respond to information will be unable to do so efficiently and effectively. Indeed, in today's information need, the main communication challenge for organisations may have shifted from how to 'unclog' blocked communication channels to how to avoid information overload (Rogers and Agarwala-Rogers, 1976: 90).

Clearly the reasons discussed earlier are not exhaustive. Organisations will undoubtedly have specific reasons related to their own operating context, culture, size and structure which will determine their ability to embrace ICT. However, many such barriers will be overcome if a good business case emerges for an investment in ICT which itself will stem from encouragement from elsewhere in the supply chain. In other words, an ICT-oriented culture will only emerge when there is a sufficient critical mass of users for its use to become the 'norm' within the industry.

Realising the potential of ICT in the future

Although references to statements such as the average modern pocket calculator having more computing power than the first lunar-lander inevitably lose their impact with repetition, the underlying fact is nonetheless valid. Moore's Law (developed by Gordon Moore in 1965) suggests that computing power will double every 18 months to 2 years (Wombat, 1997). This infers that computing power has grown at an exponential rate and it is widely believed that ICT has reached the stage where computer engineering is rapidly approaching the limits of current materials and ways of using them. The latest research programmes are examining the possibility of using light as the internal architecture of a future generation of computers. This synchronises with current developments in optical fibres technology, where the bandwidth of optical fibres is predicted to double every six months (known as Gilder's Law). This seems to be achievable through a technology known as Wavelength Division Multiplexing (WDM), which is (theoretically) capable of offering an infinite amount of bandwidth by increasing the number of wavelengths (lambdas) transferred through a single fibre. The significant aspect as far as communication is concerned is that if both Moore's and Gilder's laws continue to be applied, we could reach the situation in 2010 where communication speeds are of the order of 10,000 times higher than computation speeds (Ikeda). The implications of this are that electronic forms of communication are likely to increase as their efficiency and power enables users to overcome many of the constraints of manual methods. Thus, it is essential that the industry capitalises on this development by embracing ICT as a route to performance improvement in the future.

As was alluded to earlier in this chapter, the acceptance of ICT and realising its potential depends upon changing attitudes, cultures, structures and processes in order that it supports the way in which the organisation wishes to run its business. For example, there would be little point in implementing a new project cost control system which is totally at odds with the ways in which a firm's quantity surveyors have been used to working as the merits of such a system may be outweighed by the costs of implementation. Thus, ICTs must be designed in such a way as to be sympathetic to existing processes within the sector, rather than seek to usurp them. Another problem with communications technology is that it is often used for purposes other than for which it was developed (Poole and Desanctis, 1990). For example, telephone-answering machines were originally developed to record messages when someone was away from the telephone, but are now commonly used for screening calls before deciding whether they are to be answered (Eisenberg and Goodall, 1993: 206). This emphasises that communication technology is not neutral, but is defined by the ways in which people utilise it. Given the diversity and fragmentation of the construction industry, such problems are likely to be more acute in this respect than for other sectors because of the likely variability in utilisation. Mouritsen and Bjorn-Andersen (1991) highlight key concerns that should be considered in the analysis of communication technology. These include the fact that people are independent agents, utilise tacit as well as explicit knowledge, have informal as well as formal communication needs and undertake counter-rational decision-making. Thus, most technologies do not fail because they are technically flawed, but because their designers do not take account of the environment in which they are to be utilised.

Thus, the essence of effective ICT implementation must be to tailor the systems to the particular communication requirements of the organisation, their external business context and the needs of their employees. Key amongst the aspects of organisational life that must be considered in this process is developing a culture of acceptance of ICT innovations. Organisations will have varying views on the relative value of ICT to them. In other words, they will need to identify what are the drivers for investment in their ICT infrastructure and what will be the likely impacts on their business. The following issues are pertinent to realising the benefits of ICT within the modern construction business and should be considered as part of an ICT implementation strategy.

Developing an ICT-enabling culture

In order to realise the full benefits of ICT, particularly in providing flexibility, the organisation needs to develop a culture accepting of new technology and adapting its practices to enable its appropriate use. There also needs to be a change in the culture that underpins the *use* of such technology. McMurdo (2004) refers to the changing contexts of communication over human history. The use of tools such as those provided for communication has a shaping effect on culture and perception. However, the emphasis should be on the word 'use'; simply providing tools has no effect beyond the impact on the individuals involved in their development. Although the basic architecture of ICT systems is essentially linear and transactional (everything operates on the basis of yes/no or on/off scenarios), it is somewhat paradoxical that ICT tools actually present the possibility of creating non-linear work processes and environments. Because of the processing power of even average ICT systems, it is likely that such a system can be used to identify interdependencies that were not recognised previously. Access to data can enable the user to analyse situations in far greater depth than was previously possible, which opens up the possibility for changing job roles, adapted processes and more carefully monitored outcomes. It also raises the possibility of more IT literate employees and an organisation structure that reflects the non-linear abilities of IT to establish interdependencies.

The demands on ICT design point towards construction firms needing to adopt open system structures and developing cultures accepting of communications technologies operating at the centre of the business. In order to achieve this, the organisation must design supporting human resource systems which promote this culture shift through training and development interventions, reward strategies that recognise ICT literacy and job re-design which adapts roles to work in an ICT-enabled environment. This is as relevant at the project level as it is at the organisational level, as it will be incumbent upon the project manager to ensure that ICT systems are not circumvented by their users in favour of traditional techniques, but are adopted in a way which enhances its performance and communicative relationship with the remainder of the organisation and importantly, external parties.

Developing an appropriate structure to support ICT integration

The structure of an organisation is vital to the likely acceptance of ICT. In essence, it is vital that the structural conditions are such that the changing processes and communication practices brought about by ICT innovations can be accommodated. Rigid and hierarchical forms of organisation are such that they are unlikely to be able to accommodate changes that ICT brings about. For example, email may afford site-based workers the opportunity to communicate easily and rapidly with head office staff, but this is unlikely to work effectively if there is a protocol that all communication with head office must be relayed through the site hierarchy as a single point of communication.

One of the key lessons from history is that great enterprises of all kinds have always been underpinned by good organisation and communication (*Flexibility*, 2004a). This perspective has resulted in tasks and information being managed downwards through an organisation's hierarchy, which is built around tightly defined job descriptions and roles. This 'transactionalist' view has been reinforced through a belief that a single business location and hence, co-located workers is best, and that any new location will need a full replica of the parent organisation if it is to survive. One result of this has been the creation of information gatekeepers – individuals through whom all information flows, giving them the opportunity to dispense it when, and to whom, they deem fit. These factors can be regarded as barriers to the development of the required ICT culture and the smarter use of information and communication within an ICT context.

Two types of structure are typically identified as being able to achieve the high levels of flexibility required for effective ICT integration. First, the so-called 'organic' structure has long been identified as being more likely to survive in environments characterised by rapid change (Burns and Stalker, 1961). Organic structures come in a range of types but will consistently have fewer internal boundaries than are found in the hierarchies typical of transactional structures. Organic structures will typically remove internal and external boundaries wherever possible, but for this to work effectively, employees have to be capable of exhibiting sufficiently high levels of maturity for them to be trusted once the controlling boundaries are removed. Usually, the removal of boundaries will be a gradual process as this allows the required levels of maturity to be acquired incrementally. A second type of appropriate structure is the 'networked organisation'. This is often referred to as a delayed structure (Baines, 1998) and is possible to achieve because ICT allows for extensive horizontal and vertical communication channels to be opened up within the organisation. Baines (ibid.) asserts that the benefits of a networked organisation can only be achieved if employees are technically highly-literate, self-confident and trusted. They are empowered individuals whom the organisation trusts with access to all the information required to carry out their function effectively. Individuals who were previously supervisors now become facilitators as they seek to ease information flows rather than controlling information and thereby acting as bottlenecks in the system. They in turn have to have the maturity to accept that they can no longer rely on positional authority bestowed upon them purely because of their position within the hierarchy. Instead, they have to use their knowledge to 'earn' authority bestowed upon them by others within the organisation. This type of authority is referred to by Banner and Gagne (1995) as sapiential authority, and its existence is a strong indicator that the organisation is transforming away from being transactional.

The context of the use of ICT has thus far been predominantly that of the large organisation, but in the construction industry there is no doubt that ICT integration is as, if not more important, within the small firms that dominate the sector. An industry ICT culture can only be achieved if smaller firms are willing to be as flexible as the larger organisations with which they work when it comes to ICT adoption. Baines (1998) notes that small firms tend to be highly centralised as most, or all, of the important decisions are made by the owner or manager. In traditional organisation theory, the ability of an organisation to grow in size is constrained by the willingness of the owner or manager to allow others to make decisions. As the organisation grows, its environment becomes more complex and the quantity of information (and extent of decision-making) starts to exceed the ability of one individual. Hence, the organisation becomes increasingly de-centralised as decisions are delegated elsewhere within its structure. However, the level of decision-making support provided by ICT effectively means that it is possible to retain high levels of centralisation and still grow an organisation. Thus, in a small firm, the ICT infrastructure and the organisational structure are mutually supportive and thus, should go hand-in-hand.

Benefiting from computer mediated communication (CMC)

As a significant step towards the implementation of effective ICT, individuals, groups and organisations could consider adopting the approach referred to as computer mediated communication (CMC), a concept that has been around since the earliest days of the Internet (Kiesler *et al.*, 1984). CMC is an introductory step in the development of important ICT behaviours related to discipline and etiquette necessary to underpin its effective use within a work setting. It encourages the realisation of the trade-off between the time taken to reach a consensus, for example on action to be taken when resolving a problem, and both the equality of participation amongst the 'players' and the subsequent level of satisfaction with the decision reached. This is an important realisation because communication technologies inevitably raise the possibility of ICT overcoming the problem of equality of participation. Given equal access to suitable ICT hardware and software, large numbers of individuals may participate in decision-making in a manner similar to that of 'normal' verbal communication.

CMC could be seen as a vehicle for transforming transactional organisations (rooted in a print-based culture) into an 'electronic culture' which draws upon the benefits of ICT (Moore, 2002). It can also provide the opportunity to raise awareness of how individuals are communicating and thus provide a useful preparation for the higher level of skill, knowledge and behaviours required for successful implementation of ICT. One example is that an email can serve a number of roles, such as being a memo, a letter and a formal document. It can also, when used in a short and rapidly exchanged form, have more than a passing resemblance to a verbal communication (Kiesler *et al.*, 1984). Thus, it is important that those using this technology understand how to utilise this ICT-enabled method of communication in an appropriate manner. Evidence of the transforming nature of CMC can be found in the area of ICT development within schools, where it has been found that ICT can produce a higher commitment to learning, a sense of achievement, better self-esteem and improved behaviour (Becta, 2004). Studies have shown a range of positive impacts such as a greater ability to work independently, improved cooperative skills and enhanced confidence in communicating with others (Harris and Kington, 2002). All of these areas can be argued to contribute towards the maturity of individuals and hence of organisations.

Using ICT to promote a knowledge intensive workplace

Knowledge intensive workplaces engender an information culture, which in turn demands that those who wish to thrive within this setting develop ICT literacy. Similarly, advances in ICTs increase an organisation's interest in knowledge as it becomes a critical strategic resource (McLure *et al.*, 2000). A problem that can arise is a reluctance of organisation members to exchange knowledge with others (an example of the 'knowledge is power' mentality). McLure *et al.* (2000) examined the culture of three electronic communities of practice and found that the strongest reason why individuals chose to exchange information within a community was the desire to have access to such a community. The interpretation was that individuals were not using the community as a forum for socialising, nor to develop personal relationships, but simply valued the community as an opportunity to exchange knowledge relating to their area of practice with like-minded members. This is also good evidence of maturity and indicates that ICT-enabled communities of practice represent a useful step in the wider development of the cultural shift necessary to bring about a paradigm shift in the communication media used. Of course, this also requires the organisation to support the development of its people on this journey. It also demands widespread awareness of the IT facilities available and how they may be used as part of an integrated programme addressing facilities, people and organisation issues (*Flexibility*, 2004b).

The limitations of ICT

The preceding discussion has presented ICT as an important enabler of effective communication in organisational life. However, it is important to realise the limitations of ICT in engendering an effective communication environment. These limitations stem from the fact that people are the actors in organisation and only they can determine whether ICT will facilitate communication or merely compound the problems of information overload. Indeed, the increasing use of computer programme software may not ensure success, even at a project level. Delisle and Oslon (2004) argue that the sophistication of project management methodologies, tools and techniques does not necessarily prevent project failure. In reality, many projects fail despite tighter project controls, better methodologies and improved information flow. This is because computers do not have the ability to fully understand the functioning of the human brain, how people actually carry out actions such as separating speech from all other aural sensation or attributing meaning to each recognised word. People are also able to identify and separate factors such as tone and rhythm and use them in adding to the richness of meaning. In addition, the sender's gender, regional origins and general mood can all convey additional meaning into each piece of communication. All of this can be done without those communicating actually seeing each other, but will be determined by the innate and learned abilities of the individual to communicate and interrelate effectively.

Present day knowledge about the way in which people use multiple channels to assimilate, understand and add richness to the nature of communication have fundamental implications for the future efficacy of ICT-enabled communication processes within the construction industry. Simply providing the hardware and software that underpins ICT is not going to lead to its wholesale adoption if those using the approaches feel that it does not enhance their ability to communicate effectively. A good construction example concerns the process of resolving conflict, a bane of the industry for many decades. An amicable settlement without the need to rely on legal processes to resolve the dispute will usually require the parties to meet face-to-face and discuss their respective positions in order that a negotiated position can be reached. In these situations it is likely that ICT approaches may actually hinder the successful resolution of the problem as the richness of communication in the physical sense may be lost (see Chapter 3). Despite the undoubted advantages that ICT brings to distributed and fragmented workplaces like construction, in a people-centred industry, it is likely that ICT will never replace this fundamental interaction characteristic.

Summary

Communication technologies have the potential to change the ways in which the construction industry operates and ultimately, delivers value to its clients and other stakeholders within the supply chain. However, the use of IT in communication has been shown to present both opportunities and threats to improved communication in the construction industry. Whilst there is strong evidence that both computing power and bandwidth will continue to increase at exponential rates, there is also evidence that organisations need to consider *how* they go about using ICT in the processing of corporate information. Simply having increased computing power and functionality will not automatically result in 'better' communication. Thus, as Malone and Smith (1984) suggest, if IT is adopted in an effective manner, it will decrease an organisation's vulnerability and enhance its adaptability, but only if this process is managed appropriately.

Considering the globally recognised need to develop an enabling environment for ICT, there is a need to rethink how communication is managed through the various information technologies that are being stitched together across the sector. Although many argue that technology makes revolutionary ways of working possible, in this chapter it has been argued that it is *how* people utilise that technology that makes this possible. Clearly not all industry stakeholders have embraced the potential of ICT. Whilst this can to some extent be explained by problems in software compatibility, clear protocols are emerging for the provision of standards which allow the transfer of data between systems and organisations. Thus, it has to be concluded that the principal barriers to ICT usage within the sector are more likely to be grounded in attitudinal resistance to the changes to working that its effective use will inevitably induce. Nevertheless, given the potential of leading-edge ICT solutions to enable the industry to improve its performance, it is incumbent on all firms to foster cultures which lead to its uptake and effective use as an enabler of change.

Critical discussion questions

1 You have been approached by the owner/manager of a small subcontractor to advise on the development of an ICT strategy to exploit the opportunities with regards to E-commerce in the industry. The firm wishes to develop this strategy in order that they can expand their business and exploit new markets, particularly those in different areas of the country where they do not currently operate. Currently the firm, which employs 50 people, has no significant ICT capability beyond the occasional use of email. Devise a strategy for the firm which comprises both the technological developments necessary to move the firm into the information age and the supporting organisational development which must underpin such a paradigm shift in their operations.

2 List and discuss the key ICT challenges for a consultancy firm looking to move into international markets. In particular, consider how ICT could be used to communicate with their staff in distributed locations and what supporting structures would also need to be in place.

Case study: an innovative application of ICT in the construction process: the Stent Handheld Electronic Piling Assistant (SHERPA)

Background

Electronic communication has not really been a prominent feature on most construction sites. Most construction relies upon traditional processes and methods of capturing and transferring information between those involved. The recording of data linked to construction operations has therefore tended to be managed using manual methods such as notebooks and has therefore not been available to all of those involved in construction operations simultaneously. One example of where this situation could be improved is in the capture and use of data relating to piling operations. Unlike other construction processes, the piling process is unique in that as-built information is not often presented graphically nor can it be easily reviewed or audited as part of the project review process. Accordingly, piling specialists have traditionally relied on text-based documentation completed by the workforce. This, in turn, becomes the quality documentation providing evidence that the pile has been constructed to specification. However, the successful recording of information relies on good communications at site level and upon the accurate recording of information by the piling operatives.

The rotary bored piling process is a large and complex undertaking, allowing the construction of piles up to 3 m in diameter and 70 m in depth. It is particularly important, therefore, that a system can be developed which enables construction data to be accurately recorded at source by the operatives in order that it can be utilised by others involved in piling operations both on-site and in office locations. This case study explains the development of an innovative ICT-based system for capturing and utilising data connected with piling operations.

The communication challenge in capturing and communicating piling operations data

During major piling operations, a significant amount of information and data must be communicated between the various parties on the site. Information needing to flow between project participants includes:

- pile construction schedules containing levels, diameters and rebar details;
- specified design tolerances;
- site levels;
- on-site calculations for target depths and set levels using design and site data;
- site volumes and quantities of materials delivered and installed; and
- as completed pile records.

Although some of the data is derived from design information, much originates from the working gangs themselves. This presents a significant challenge in terms of capturing data and recording it efficiently. Between the design team and the piling operatives, a range of other participants must relay and act upon this information including foremen, suppliers, site engineers, design engineers and the client. Thus, the information flow associated with Rotary Bored Piling process must transcend many interfaces if design and piling information is to be recorded and acted upon successfully.

An analysis of site piling processes revealed that the quality of communication at the site level was highly variable. Observation of the piling process showed that many factors affect the quality of communication during piling operations. The key factors identified included:

- The problems inherent in paper-based pile design schedules and data recording which increased the possibility of errors in the process.
- Late design data being delivered to site by the client which often supersedes that which the operatives are working to.
- Verbal communication of important data between working gangs which can lead to transcription errors and which leaves no archive of the decisions made in relation to the piling process.
- Errors in manual calculations which are usually contained within the foreman's notebook.

- Unstructured data stemming from disparate forms of recording as-built data.
- A lack of good quality production data which can lead to unobtainable target outputs.

The problems observed in the piling process were found to have both direct project-specific impacts for piling operations on site, as well as indirect implications for the efficiency of data management for the organisation involved. On site, poor communication and data capture was found to lead to poor quality and out of tolerance piles being constructed. In extreme cases, it had even led to some piles not being constructed and hence, to considerable rework, additional costs and severe delays to the contract. A lack of knowledge regarding the current status and progress of piling operations at head office had also led to problems remaining unresolved that could have been dealt with by senior engineering staff. At a company wide level, an inability to capture adequate data for re-use within the head office had detrimentally affected processes such as pile design and estimating. Having a ready source of on-site design and construction data would provide an archive of information for informing future piling operations. Thus, the problems inherent in traditional communication processes around the piling process were such that it presented a clear case for the use of information technology to facilitate data capture and handling.

The SHERPA system

The aim of the SHERPA system was to develop an electronic site-based wireless network for the collation of pile schedules, design specifications and as-constructed data. The network is managed by a site-based web-server with a user interface comprising a number of user operated touch-screen tablet computers (the number of workstations being determined by the extent of the piling operations – see representation in Figure 8.1). These are operated directly by the site operatives allowing them to enter and record data, carry out piling operation calculations and resolve difficulties in real time. The operation of the system is managed as follows:

- Design data, specifications, schedules, tolerances and contract settings are loaded into the site server at the outset of the contract and maintained as a single data source throughout the contract. Changes to design can be automatically updated via the central server and communicated to the remote network of site-based work stations.
- Users collect data using a series of process-orientated web-based data capture pages, data from which is stored within the site server. No data is stored directly within the tablet but is immediately transferred through the wireless network.

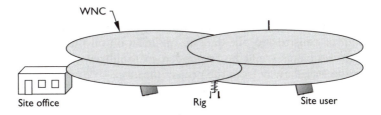

WNC

Site office   Rig   Site user

Figure 8.1 Typical wireless network cell configuration.

- As the piling data is captured it is verified against contract tolerances and then users are immediately alerted to possible problems such as incorrectly specified piles or non-conformance to contract requirements. As errors are corrected they are entered into the remote workstations and the system automatically updates the as-built information.
- All calculations and volume checks are carried out automatically by the SHERPA system, thereby eliminating the possibility of site calculation errors.
- Users can access any data page for any piling operation. This allows the greatest possible flexibility for all users, especially on very large projects where various piling operations can occur concurrently.

The use and benefits of the SHERPA system

The SHERPA system has revolutionised the way in which piling data is captured, communicated and reconciled with design data. Direct benefits observed from implementation of the system include a marked reduction in defective work derived from piling operations. Mistakes induced by miscalculations or incorrectly specified and communicated operations have been virtually eradicated by the system leading to significant cost savings in avoidance of rework. An example of this is the significant reduction in the cost of concrete that has been achieved, as the required volume is calculated by the SHERPA system removing the need for over-ordering as a contingency measure. In addition, improving the access to data for all of those involved in the process and regulating the data structure has provided significant benefits for future re-use in estimating and planning operations. In addition to improving the management of information associated with the piling process, SHERPA has yielded a range of additional benefits which were not necessarily envisaged at the time of its development. For example, the system encouraged the active involvement of site-based operatives undertaking the manual work who hitherto had no data collection responsibilities.

The system also overcame the variability in working patterns of site staff caused by plant breakdowns, accessibility problems, supplier delivery and unforeseen ground conditions. By allowing all site users the ability to access any design or constructed pile data at any location on the site and at any time, these problems could be responded to rapidly without recourse to other parties based away from the project location.

Enabling technologies underpinning the SHERPA system

The SHERPA system combines state-of-the-art hardware and software systems to provide a practicable system of data capture within the site-based environment. Two important features of the system developed that have made its achievements possible are its web-based interface and the totally wire free network on which it is based. The web-based system allows for net-distributed applications to be developed allowing access by all users both on site and off-site using the same interface. The platform independence offered by web-based applications means that all that is required is a web browser on any personal computer and so access can be obtained from anywhere in the world. This offers particular advantages for distributed design teams (i.e. virtual teams) who are increasingly based remotely from the production location. Coupling the web-based interface to a wireless local area network (WLAN) has allowed a totally unwired system to be developed enabling total flexibility in positioning the tablet computers around the site. This was necessary as the on-site operations require that the tablets were located such to avoid damage and conflict with ongoing works. The WLAN is packaged in crush-proof casings that can be standalone, attached to piling rigs or other site plant as appropriate. Power is provided by 12 V battery or directly from the construction plant allowing the WLAN to be easily reconfigured as necessitated by the work or diverted around objects which may impede the signal.

Challenges in implementing and benefiting from SHERPA

Although the SHERPA system can be seen to offer significant benefits to on-site piling operations, there remain a number of people-related barriers that must be overcome if its benefits in improving the communication process regarding piling data are to be fully realised. One of the central problems concerns the variable IT literacy levels of the site-based staff using the system. This problem is compounded by the high level of temporary and agency staff who may not be using the system from one site to another. This demands that training is provided in-situ to enable users to grasp the fundamentals of the system and use it in the way intended. In this respect, the abilities of supervisors to encourage the use of the system and to train users in its effective operation is crucial. Another challenge concerns the ways in which the data captured by the SHERPA system is utilised in both on-site and head office/future project operations. The level of data is far beyond that which has been able to be captured before. Its utility in shaping the success of future operations depends upon the availability of time to query and handle the data by the engineers.

Conclusions and lessons learned

This case study has provided an excellent example of how the use of advanced ICT has been implemented in such a way as to enhance, rather than usurp, the work of the end user. The SHERPA system has revolutionised the way in which piling data is captured and utilised on site. It is flexible enough to account for the variability in project size, the number of personnel on the site and different specifications for data collection imposed by the client. The system is totally flexible, easily reconfigurable expandable and scalable. However, it has also yielded significant benefits from the perspective of the site workforce in supporting their understanding of the piling processes. The operatives have fully engaged with the system and have gained an appreciation of the importance of accurate data capture. The system has also facilitated accessibility to the data from the perspective of the client and their advisors. They can review real-time information on the performance and quality of piling operations, reassuring the design team that specifications have been met and that the finished product conforms to the design specification. Thus, the case study has demonstrated the importance of developing a robust knowledge of the processes that it is aimed to support, and the needs of those expected to use it, in order for such a system to be accepted and utilised effectively.

The SHERPA system was the winner of the *Effective IT Awards 2004* in the category of '*Most Effective Use of Mobile and Wireless*'. Further information on the SHERPA system can be obtained from Michael Ward at MJW@Adtec.uk.com

Chapter 9

Conclusions and future directions in construction communication

Introduction

This book has explored the nature of project-based working in the construction sector and the reliance of the industry on effective communication at all levels of its operation. It has highlighted the inexorable link between effective communication and the performance of construction projects, highlighting the factors that can impede open communications within the industry and how such barriers can be overcome. Considering that construction is such a fragmented and dynamic sector, the challenges of communicating effectively are manifestly greater than in most other production environments. Contractually driven relationships, conflict and a lack of mutual respect and trust, all combine to compound what is a complex structural communication environment. These factors, in turn, render the role of the project manager as a communication facilitator extremely demanding and stressful. Nevertheless, improving communication in the industry can be seen as the principal enabler for improving its performance in the future. This concluding chapter aims to bring together the themes that have emerged as being salient to the context of the construction industry within the body of the text and to provide some thoughts and directions as regards the future of communication in the sector.

The construction communication context

As has been explained throughout this text, the complexity and dynamism of the industry's project-based structure and culture threaten to undermine the applicability of many of the central tenets of effective communication practice that have been applied successfully in other sectors. Unfortunately, many of the management practices that have evolved in response to the unique context of the sector have done little to engender an open communication environment that ensures conjoined team working, process integration and improved performance. It has been revealed that, despite radical improvements in ICT capability, there remains a reliance on formal face-to-face exchanges that take place between industry stakeholders. In other words, construction remains a 'people industry' and it is within the success or otherwise of this human interaction that the future of the industry is firmly rooted.

This book has provided a contextual backdrop to the construction industry by way of explaining why communication is so problematic in comparison to other sectors. It is important in concluding this book that some of these problems are revisited and potential solutions to them identified based on the theoretical and practical approaches explored within this text. First, construction is a project-based industry. In other words, people are brought together for short periods to work together in a temporary endeavour, before moving on to other ventures. Thus, the communication structures which stem from this temporal involvement climate must evolve rapidly, and be supported by a degree of common understanding amongst the participants, if they are to be successful. This common understanding forms a component aspect of the culture of the industry, whereby participants have grown used to a set of communication norms which centre on human interaction as the norm. Whilst it may seem relatively unsurprising for such a labour intensive industry involving such a large number of people, with so many individual interactions occurring simultaneously and across so many organisational interfaces, it is little wonder that problems occur. Second, the construction industry is volatile, dynamic and is subject to rapid change and fluctuation. At a macro level, construction organisations' fortunes are inexorably linked to the prevailing economic circumstances. Considering that capital investment is often the first activity that the industry's clients suspend at times of economic downturn, but one of the prime areas for investment in times of economic growth, it is little wonder that the sector finds itself lurching from peak to trough in terms of both output and demand. In the latter part of the twentieth century, this resulted in most construction organisations developing into 'hollowed out' firms which directly employ few staff, preferring instead to outsource as much of their labour as possible. This has provided construction companies with a high degree of flexibility in terms of coping with variability in demand, but the inevitable implication has been to create a degree of temporal involvement in organisations as well as projects. This has bred unfamiliarity and poor communication at both an intra-organisational and inter-organisational level. Add to this the sheer number of different people involved in construction activities and it can be understood why the structure of the industry renders communication even more problematic than in other project-based industries.

For many years, poor communication practices have been recognised as a serious delimiting factor within the construction industry's efforts to improve its performance and the satisfaction of its client base. Within the United Kingdom, a succession of government-commissioned reports has berated the industry for its apparent inability to communicate effectively, both internally and externally. Over the past two decades, an abundance of management initiatives have expounded more effective communications mechanisms as a route to the idealised concept of the 'post-modern organisation'. However, there is a paucity of evidence as to the take-up of such approaches within the construction industry. Rather, the sector appears to have ignored many of the espoused communication panaceas for the 'knowledge-based' economy.

So what are the factors that prevent the industry working together more effectively and achieving the step change improvement in performance demanded by its client base? This is a difficult question to answer precisely because the industry is so large, complex and diverse. Everyone involved in the industry plays his/her part in an intricate communication network. Seeing the project environment as an interconnected network of actors is appropriate because every such venture, no matter how small or well defined, cannot be successfully completed without interactions and transactions between people and organisations. Unfortunately, from a communications perspective, people working within construction cannot be relied upon to act and interact in a uniform/standard manner because they came from diverse backgrounds with different perspectives and will thus have dissimilar needs from their interactions with others. Indeed, as was alluded in Chapter 1, it is precisely *because* of people's idiosyncrasies that construction presents such a fascinating environment within which to explore communication processes and practices.

Deconstructing the construction communication challenge for the project manager

The approach adopted within this book has been to break down communication within the industry into different levels or contexts in order both to facilitate understanding and to provide a framework from which to address the needs of the sector. The book began by exploring the nature of interpersonal interaction amongst project stakeholders. At the level of personal interaction, many factors can impinge on effective communication. For example, different industry participants communicate in different languages which relate to their own disciplines. Furthermore, a range of non-verbal cues affect the message being relayed involved and a variety of different types of noise (in both real and metaphorical senses) surround the interactions of those involved. At a group or team level, the communications processes and the nature of interaction becomes even more complex. The way in which people work together is affected by a wide range of factors stemming from the temporary involvement climate. Developing high performance teams demands that they are managed in a way which exploits group synergies whilst maintaining a degree of order and coherence to their work. This process appears even more complex when considered in the context of inter-organisational working, where cross-company working introduces the possibility of differing objectives and external conflict within the team's interaction. Finally, the issue of corporate and mass communication raises a completely different set of effective communication criteria which must be addressed. Understanding how large groups of people will react to communications processes demands a different set of communication methods and competencies.

When it is considered that many managers within the industry communicate across all of these levels simultaneously, the complexity of the project management role can be fully appreciated. Consider the project manager who may in the course of a day communicate with individual colleagues, with their internal team, the wider group of project stakeholders and even other stakeholders external to the formal project organisation. They must continually draw upon a range of different communications approaches to ensure that the right messages are communicated to those who require them. It can be seen that success as a construction project manager is in many ways defined by their ability to communicate effectively at all of these levels.

Future construction communication challenges

Throughout this book, the challenges facing the industry have been explained and their impacts on communication within the sector discussed. In this concluding chapter, it is important to highlight those which are likely to continue to pose problems for those working in the industry and to identify some of the approaches that might begin to address their impact.

The performance improvement imperative

In recent years, much has been made of the industry's need to change and respond to increasing client demands for improvements in the delivery of the sector's projects. There has been a steady increase in the quality of service and product expected by clients procuring construction work. Inevitably, improving performance demands more effective ways of communicating between parties in order that client needs are interpreted and their expectations managed in a manner that accounts for the natural constraints on performance that the industry provides. Although it is certainly easier to communicate more effectively where long-term relationships are in place (which are certainly becoming more prevalent since the growth in partnering and new procurement regimes), these new processes and practices will themselves demand new and more effective ways to communicate, breaking down traditional boundaries and overcoming process discontinuities.

The drive for integration

Defects and rework in construction can be attributed to design errors, materials failure, workmanship problems or to discontinuities in the design and construction process. Communication problems can be seen to be a contributing factor in all of these. In order to overcome such failings, recent reports have emphasised the importance of integrating design and construction processes and importantly, the various members of the supply chains involved. However, if such integration is to be meaningful in terms of joining-up construction activities, effective communication channels must be developed which enable those involved to overcome ingrained functional boundaries and culturally defined practices which have prevented integrated working in the past.

The shift towards a service-oriented industry

Despite its production focus, construction is not immune from the economic shift towards a service-sector-oriented economy. Most construction companies now operate as 'hollowed out' firms; whereas in the past they would have directly employers operatives and craftspeople, most now comprise a group of core managers with the remainder of their labour needs being met through outsourcing. This presents additional challenges for construction firms who have to procure additional services and productive capacity from *external* suppliers. Every new external interface presents the potential for additional problems in the communication process and presents new boundaries which have to be managed. Given the nature of the industry's operation, these can differ from project to project and so have the tendency to create new challenges for every new endeavour.

The need to influence workforce behaviours

Organisational communication is arguably a key factor in embedding new ways of working necessary for change within construction organisations. Communication is not merely a mechanism to convey or transmit information, but is a tool by which workforce attitudes and behaviours can be changed and manipulated. In recent years, some construction companies have begun to recognise the power of softer behavioural skills competencies in defining the success of an organisation (see Cheng *et al.*, 2005). This acknowledgement stems from the realisation that it is the behavioural input to a project's development that determines its success. Such behaviours are manifested and conveyed to others in the ways in which project participants communicate. Thus, training, developing and supporting people in improving their communication skills is central to the improved performance of the sector in the future.

The need to embrace workforce diversification

Multiculturalism is an increasingly prominent feature of the modern construction industry in many countries. Indeed, even in the homogenous UK construction industry, pressure is growing for the sector to start attracting more women, ethnic minorities and other groups which have been historically underrepresented. This will impact upon the ways in which people communicate in the sector. Whereas in the past reliance on traditional male-oriented ways of communicating may have been considered completely acceptable and appropriate, this will not remain the case in the future. There will be a requirement for those working in the industry to be sensitive to the communication needs of others. Similarly, EU accession is leading to an influx of foreign workers to the sector who are arguably desperately needed to offset the skills shortage brought about by the industry's low level image. People from disparate cultural and social backgrounds may interpret different meanings into the communications that they have. Ensuring that this does not inhibit collaborative working and performance is bound to present a major challenge to construction companies in the future. Effective communication lies at the heart of being able to benefit from the advantages that workforce diversification potentially offers.

Globalisation

The onset of global construction markets has recently begun to be felt within the sector at large. Fast growing economies such as the Chinese construction industry offer massive market potential for European companies willing to invest in attempting to exploit them. However, taking advantage of new markets presents serious communication challenges. These include, for example, the difficulties of providing the language skills and cultural knowledge necessary for their employees to be able to communicate in unfamiliar environments. It also demands robust communication paths for those managing projects in different countries to maintain contact with their employing organisation. The need to adapt to competitive and dynamic markets is therefore proving a popular theme in contemporary organisational communication studies.

New communications technologies

Whilst on one hand, modern information and communication technologies are revolutionising communication within construction, they have also changed the way in which people interact, reducing the face-to-face contact that can overcome many of the relational problems known to beset the industry. Thus, whilst the onset of electronic commerce and web-based technology is an exciting prospect for the sector, their adoption must be treated with discretion and caution if their use is not to undermine the human communication and interaction upon which the industry's historical evolution has been founded. Understanding the impact of technology on the communication dynamics of the sector will therefore be crucial in the future, as will gaining acceptance of ICT-enabled communication, particularly amongst the less IT literate.

The need to reduce 'noise' and overload in the construction communication process

'Noise' is important because it can impair the understanding of the message by the receiver, no matter how appropriate the channels and media that are selected. In construction, noise problems can arise because of the physical environment of the construction site, the number of interfaces in the message chain and the emotional impact of those involved in the industry. Overcoming noise in all its forms must be a key priority for the sector if its communication processes are to be improved. A related problem is the need to avoid information 'overload' in today's communication age. Within the construction industry, a plethora of different communication types can occur concurrently which has the potential to detrimentally affect the abilities of people to process information effectively. Furthermore, a proliferation in ICTs within the industry is creating the possibility of those involved being able to retrieve massive amounts of data which they may be ill-equipped to manage. Helping those that work in the industry to identify the information that is important to them and discard the superfluous is therefore crucial to integrating ICT within the sector in the future.

The need to communicate a new image for construction

The low level image of the construction industry in many countries and sectors has the potential to delimit its appeal to the high achievers it needs to attract, as well as undermining trust in its organisations on the part of clients and other stakeholders. Considering that the industry relies on just about every other sector for its workload, generating a more positive and professional image for itself should form a key driver for its development in the future. Promotional campaigns to improve the image of the sector have become an established activity across the industry. For example, a National Construction Week campaign (National Construction Week, 2004) is run every year, which provides young people with the opportunity to experience the wide range of opportunities available within the industry. However, it is incumbent upon all of those with a stake in the sector, from construction operatives to leading client bodies, to contribute to communicating a better image for the industry amongst the general public if its future is to be safeguarded. Communicating a consistent message will be particularly problematic given the sheer diversity of the sector and those that work within it.

The role of communication in enabling change

The construction industry is arguably going through a period of irrevocable change. In response to exhortations from government, leading clients and major firms, the sector is attempting to move to a stage where adversarial working and poor relations between project participants will not be tolerated and where performance improvements are delivered through effective collaboration between the parties involved. Despite considerable efforts in this regard in recent years, the industry clearly has a long way to go before this espoused aim is achieved. The disconnection between the integration required to deliver the industry's 'performance improvement agenda' and the fragmented and disparate reality of the sector, remain significant barriers which are yet to be overcome. However, effective communications offer a potential conduit to delivering on the performance outcomes expected. Indeed, the implementation of any change and transition process demands effective communication processes to support it (Weiss, 2000: 213). It is somewhat surprising, therefore, that more attention has not been paid to defining the communication structures necessary to support the desired change. Rather, the expectation appears to have been for effective communication to emerge as a corollary of new ways of working. A key theme running through this book has been that defining effective communication structures should form a *precursor* of any attempt to change the ways in which the industry operates.

The extremely problematic context presented above demands that construction organisations prioritise the development of a robust and flexible communication strategy that recognises both internal communication needs, but also the prevalent culture of both the wider organisation and the individual contract. For this reason, larger- and longer-term projects will often define their own communication systems and protocols, or will allow them to evolve during the duration of the project. This view of planning communication strategies reflects a contingency view of organisational design and recognises the need to build on the strengths of the individuals involved in projects. It also rejects the notion of applying a 'one-size-fits-all' communication policy across many projects. Such normative communication processes are highly unlikely to address the issues raised within this book. Indeed, the perspective adopted in this text has been to reject the notion of 'good communication theory', which broadly assumes that the aims of managers and employees are the same, that differences in opinion stem from misunderstandings and that the solution to industrial relations problems is to improve communications (Armstrong, 2001: 808). Indeed, it is easy to criticise such a simplistic view of organisational life considering the divided loyalties which are bound to exist within and between temporary project organisations and their constituent members.

Ignoring the exhortations to adopt 'best practice' approaches to improving the industry's performance in favour of tailored and bespoke communication model demands that those involved in construction projects take time to understand their communication needs. As was alluded to above, every individual project endeavour will involve a different set of stakeholders, each of whom will have their own communication requirements which may or may not align with that of the other project players'. Identifying potential cultural and organisational barriers to inter- and intra-project communication and designing systems (and where necessary protocols and procedures) for overcoming them will be time well spent in avoiding the problems known to be detrimental to collaborative working and ultimately, project performance. However, using communication as a route to improve outturn performance should not form the sole driver for putting effective communication structures in place; an effective communication environment leads to a fairer, more open and inherently more satisfying workplace environment for all involved. Persuading project stakeholders to take time out to understand how their individual project communication can be improved remains an aspiration in most cases, but is arguably crucial if significant improvement in the ways in which people work and operate is to be achieved.

The construction project manager as an enabler of communication

Whilst this book has sought to raise the reader's awareness of the unique context in which communication takes place within the construction industry and the implications that this inevitably has on the effectiveness of the sector, a concurrent focus has been on identifying the crucial role of the project manager in facilitating such processes and methods that they can use to fulfil this role more effectively. Thus, it is appropriate in concluding this text, that consideration is given to how the project manager can enable effective communication within the sector. The construction industry relies on the abilities and skills of line managers to a greater extent than in most other sectors. Project managers must ensure that their teams are motivated, working towards a common goal *and* communicating effectively and coherently. They are expected to articulate this vision in such a way that their teams, who are likely to be a group of disparate individuals who may have never worked together before, synergistically combine their skills and knowledge effectively towards the task. It is clear, therefore, that in order to be effective in bringing together the project team, a project manager must be able to communicate at all levels, from the project sponsors down to the most junior members of their team.

Interaction with different project stakeholders and team members requires different communication abilities. A project manager's competency in encoding and decoding therefore plays a crucial role in achieving project outcomes; if they are effective in communicating their thoughts through the encoding and decoding process, they are more likely to achieve their desired outcomes and hence, be more successful in their role. These communication abilities are not innate, but must be developed by managers in these demanding roles. Communicating effectively with a client's representatives and subcontractors arguably demands very different sets of skills and radically different approaches, but all project managers must learn to tailor their approach to the circumstances that present themselves.

The demanding and multifaceted nature of the project management role emphasises the contention discussed at the outset of this book, that there are no generic methods of communication which can applied to all situations. Rather, the project manager must develop a communication style that accords with the particular project or situation at hand. In other words, the techniques used should be contingent upon the particular circumstances to which they are to be applied. Adopting a contingency view to understanding communication recognises the impact that environmental variables have on the decision of how to communicate most effectively.

Some future research directions

Although this book is not necessarily aimed at researchers or the research community, it remains important to recommend some indication of potential future research directions given the complexity of communication in the construction industry and the relative lack of focus on the issue to date. These are areas which are likely to continue to pose problems for those working in the industry and to identify the types of approaches that might begin to address their impact:

- *Communication as a route to individual and organisational learning* – a significant research opportunity concerns establishing the role of communication in transforming construction firms into learning organisations. Effective communication lies at the heart of this process in acting as the enabler of knowledge diffusion, but ways of achieving this in dynamic, project-based environments remain poorly defined.
- *Communication as a route to effective HRM* – communication is probably the most important enabler of effective human resource management but, when of poor quality, it also has the potential to severely limit their effectiveness. HRM communications must have both internal and external dimensions. The internal dimension must focus

on ensuring effective communications between managers and workers in different parts of an organisation, particularly project staff and central HRM departments. In contrast, the external dimension should focus on communications with external interest groups such as governments, pressure groups, local communities and trades unions. Both internal and external communications are under-researched in relation to effective people management, but new knowledge in this area is vital if high levels of commitment are to be achieved.

- *Communication as a route to trust and collaborative working* – effective communication is a prerequisite for effective working relationships founded on trust and mutual understanding. Collaborative and integrated working cannot occur, therefore, without the development of open channels of communication between all parties in the supply chain. Research is required to refine these channels and to establish the necessary conditions upon which successful communication can be founded.

Although these areas are all important to the future of the industry, the list is by no means exhaustive. Thus, it is recommended that the reader generates his/her own lines of enquiry aimed at addressing problems with which he/she is familiar. Indeed, given the individualised nature of communications networks which emerge in virtually every construction project, each will provide a unique arena within which to undertake research.

Summary

Rather than viewing effective communication as an important facet of project management, in this book it has been viewed as an essential *prerequisite* of successful project management; if we communicate more effectively, then other managerial processes should work more effectively as a result. However, rather than prescribe a set of guiding principles for managers to adhere to and thereby improve communication performance, it has been recognised that there is no single communication paradigm or panacea for an industry as complex as construction. This complexity demands that managers tailor their approaches to the project at hand. The reader should apply the knowledge and principles contained within this text to identify the particular communication needs of a given situation. In this way they can establish more effective ways of identifying and responding to their own communication challenges and the context within which they are manifested.

Critical discussion questions

1 Discuss the communication challenges brought about by new forms of procurement such as the Private Finance Initiative (PFI) projects.

Evaluate how the project manager must adapt their communication style to accord with the demands of such projects.

2   List and discuss three areas which demand fresh research to improve communications within the industry. Design a research methodology to explore these problems drawing on the principles learnt within this book.

References

Agapiou, A. (2003) 'Exploring the attitudes of school-age children, parents and educators to career prospects in the Scottish construction industry', in Proceedings of 19th Annual ARCOM Conference, 3–5 September, University of Brighton, 1: 243–51.

Ahuja, M.K. and Carley, K.M. (1998) 'Network structure in virtual organisation'. *Journal of Computer Mediated Communication.* 3(4): 1–31. [Online]http://www.ascusc.org/jcmc/vol3/issue4/ahuja.html

Albrecht, T. and Hall, B. (1991) 'Facilitating talk about new ideas: the role of personal relationships in organizational innovation'. *Communication Monographers,* 58: 273–88.

Alshawi, M. and Underwood, J. (1999) *The Application of Information Technology in the Management of Construction.* RICS Research, London.

Ambrosio, J. (2000) 'Knowledge management mistakes'. *Computerworld.* 34(27): 44–5.

Anumba, C.J. and Evbuomwan, N.F.O. (1997) 'Concurrent engineering in design-build projects'. *Construction Management and Economics.* 15: 271–81.

Aouad, G. and Kawooya, A.A.O. (1996) 'Construction planning methods: a review of history, capabilities and limitations'. *Journal of Construction Procurement.* 2(2): 19–37.

APM, *Association for Project Management* (2000) *Project Management Body of Knowledge* (4th Edn), Dixon, M. (ed.), Association for Project Management, Buckinghamshire.

Armstrong, M. (2001) *A Handbook of Human Resource Management Practice* (8th Edn), Kogan Page, London.

Atkinson, A.S. (2002) 'Ethics in financial reporting and the corporate communication professional'. *Corporate Communications: An International Journal.* 7(4): 212–18.

Audit Scotland (2002) The 2001/ 02 Audit of the Scottish Parliamentary Corporate Body, Additional report on the Flour City Contract, Scottish Parliamentary Corporate Body.

Axley, S. (1984) 'Managerial and organizational communication in terms of the conduit metaphor'. *Academy of Management Review.* 9: 428–37.

Baguley, P. (1994) *Effective Communication for Modern Businesses,* McGraw-Hill, London.

Baines, A. (1998) 'Using information technology to facilitate organisational change'. *Work Study*. MCB University Press, Bradford.

Bale, J. (2001) Quoted in *Construction News*, 'Image fails the Test' 1 March, pp. 12–13.

Bales, R.F. (1950) *Interaction Process Analysis: A Method for the Study of Small Groups*, Addison-Wesley Press, Cambridge, MA.

Balmer, J.M.T. and Gray, E.R. (1999) 'Corporate identity and corporate communications: creating a competitive advantage'. *Corporate Communications: An International Journal*. 4(4): 171–6.

Balmer, J.M.T. and Gray, E.R. (2003) 'Corporate brands; what are they? what of them?'. *European Journal of Marketing*. 37(7/8): 972–97.

Banner, D.K. and Gagne, T.E. (1995) *Designing Effective Organisations*, Sage Publications, Thousand Oaks, CA.

Banwell, H. (1964) The Placing and Management of Contracts for Building and Civil Engineering Work, HMSO, London.

Barnard, C. (1938) *The Functions of the Executive*, Harvard University Press, Cambridge, MA.

Barthorpe, S. (1999) 'Considerate contracting-altruism or competitive advantage?', in *Profitable Partnering in Construction Procurement*, S.O. Ogunlana (ed.), E&FN Spon, London, pp. 179–88.

Beattie, V., McInnes, B. and Fearnley, S. (2002) *Narrative Reporting by Listed UK Companies: A Comparative Within Sector Topic Analysis*, University of Stirling, Stirling, www.stir.ac.uk/Departments/Management/Accountancy/stfpages/beattie/ spring%20paper%20final%20.pdf (accessed 16/01/05).

Beaumont, P.B. and Hunter, L.C. (2003) *Information and Consultation: From Compliance to Performance*, Chartered Institute of Personnel and Development, London.

Becta (2004) *What the Research says about ICT and Motivation*. www.becta.org.uk/ (accessed 05/08/05).

Belbin, M. (2004) www.belbin.com/belbin-team-roles.htm (accessed 22/06/04).

Bella, D.A. (1987) 'Organisations and systematic distortion of information'. *Journal of Professional Issues in Engineering*. 113(4): 360–70.

BICRP (1966) *Archives of the Building Industry Communications Research Project*, RIBA, London.

Birrell, G.S. and McGarry, R.G. (1991) 'The building procurement roles – their professional personalities and communication issues among them', in *Management, Quality and Economics in Building*, Bezelga, Artur and Brandon, Peter (eds), E&FN Spon, London, pp. 43–52.

Black, R. (2000) The New Scottish parliament Building: An Examination of the Management of the Holyrood Project, Audit Scotland, September.

Black, R. (2004) Management of the Holyrood Building Project, A Report prepared for the Auditor General for Scotland.

Blackler, F. (1995) 'Knowledge, knowledge work and organisations: an overview and interpretation'. *Organisation Studies*. 16(6): 1021–46.

Blockley, D. and Godfrey, P. (2000) *Doing it Differently: Systems for Rethinking Construction*, Thomas Telford, London.

Boddy, D. and Paton, R. (2004) 'Responding to competing narratives: lessons for project managers'. *International Journal of Project Management*. 22: 225–33.

Bodensteiner, W.D. (1970) *Information channel utilization under varying research and development project conditions: an aspect of inter-organizational communication channel usage*. PhD Thesis, The University of Texas, Austin, TX.

Boudjabeur, S. and Skitmore, M. (1996) 'Factors affecting performance of design build projects', in Proceedings of ARCOM 12th Conference, 11–13 September, Sheffield, UK, pp. 101–10.

Bowden, S. and Thorpe, A. (2002) 'Mobile communications for on-site collaboration', in Proceedings of ICE, Civil Engineering, 150, November, Paper 12989, pp. 38–44.

Boyd, D. and Wild, A. (2003) 'Tavistock studies into the building industry: communications in the building industry (1965) and interdependence and uncertainty (1966)', in *Construction Reports 1944–98*, Murray, M. and Langford, D. (eds), Blackwell Publishing, Oxford, pp. 69–85.

Bragg, T. (1999) 'Turn around an ineffective team'. *IEE Solutions*. 31(5): 49–51.

Brown, S.A. (2001) *Communication in the Design Process*, Spon Press, London.

Bryman, A., Bresnen, M., Beardsworth, A.D., Ford, J. and Keil, E.T. (1987) 'The concept of the temporary system: the case of the construction project'. *Research in the Sociology of Organizations*. 5: 253–83.

Building (1997) 'Big Changes in Store at Tesco'. 5 December, pp. 14–15.

Building (2001a) 'Model behaviour'. 2 February, pp. 47–51.

Building (2001b) 'Cracking down on drugs'. 29 June, pp. 25–6.

Building (2001c) 'Image is everything'. 4 May, pp. 52–4.

Building (2001d) 'World wide winners'. 5 October, pp. 49–50.

Building (2002a) 'Mind your language'. 1 March, pp. 24–5.

Building (2002b) 'Why are we still so clueless?', in *Supplement: Rethinking Construction: Your Guide to the Egan Revolution*, November, pp. 18–20.

Building Careers Supplement (1999) 'Why waste half your life working in all that noise and dust? It's just not worth it'. October, pp. 10–12.

Bunn, R. (1997) 'Shrinking distance'. *Building Services Journal*. 19(3): 29–31.

Burns, T. amd Stalker, G. (1961) *The Management of innovation*, Tavistock Publications Ltd, London.

CALIBRE (2000) www.calibre2000.com/# (accessed 10/12/04).

Campbell, D.J. (2000) 'Legitimacy theory or managerial reality construction? Corporate social disclosure in Marks and Spencer plc corporate reports, 1969–1997'. *Accounting Forum*. 24(1): 80–100.

Cantu, C. (1998) 'Virtual Teams'. CWTS Report. University of North Texas, *Implementing Groupware Technology in Organizations for Facilitating Distributed Teamwork*, in Thompson, A.R., www.netspace.org. (accessed 30/08/05)

Carlsson, B., Josephson, P.E. and Larson, B. (2001) 'Communication in building projects: empirical results and future needs', in Proceedings of CIB World Building Congress: Performance in Product and Practice, 2–6 April, Wellington, New Zealand, Paper HPT 29 (CD copy).

Carr, B.P. and Winch, G.M. (1999) Measuring on-site Performance in Britain and France: A Calibre Approach, Bartlett research paper, No.9.

Carter, C. and Scarbrough, H. (2001) 'Towards a second generation of KM? The people management challenge'. *Education + Training*. 43(4/5): 215–24.

Cheng, E.W.L., Li, H., Love, P.E.D. and Irani, Z. (2001) 'Network communication in the construction industry'. *Corporate Communications: An International Journal.* 6(2): 61–70.

Cheng, M.-I., Dainty, A.R.J. and Moore, D.R. (2005) 'What makes a good project manager?'. *Human Resource Management Journal.* 15(1): 25–37.

Christensen, L.T. (2002) 'Corporate Communication: the challenge of transparency'. *Corporate Communications: An International Journal.* 7(3): 162–8.

Church, A.H. (1996) 'Giving your organization communication C-P-R'. *Leadership and Organizational Development Journal.* 17(7): 4–11.

CICA (1993) *Building on IT for Quality.* Construction Industry Computing Association, Cambridge, MA.

CITB (1998) *Children's Attitudes Towards the Construction Industry – A Research Study Among 11–16 Year Olds*, CITB, King's Lynn.

CITB (2004) *Skills Foresight*, CITB, King's Lynn.

Clevenger, T. Jr. and Matthews, J. (1971) *The Speech Communication Process*, Scott Foresman, Glenview, IL.

CNAEC (2004) *China Yangtze Three Gorges Project Development Corporation (CTGPC)*, China National Association of Engineering Consultants, Beijing. www.cnaec.org.cn/sponsors/Sponsor_CTGPC.htm (accessed 05/08/05).

Connaughton, J. (1993) *Investment divisions in large capital projects.* Unpublished PhD thesis, University of Greenwich, London.

Conrath, W.D. (1973) 'Communication patterns, organizational structures, and man: some relationships'. *Human Factors.* 15(5): 459–70.

Considerate Constructors Scheme (2004) www.ccscheme.org.uk (accessed 23/06/04).

Constructing Excellence (2004a) Innovation 1 – iTAG – electronic tagging of materials and components, Bovis applies Safeway thinking www.constructingexcellence. org.uk/regions/yorkshirehumberside/details.jsp?innID=59&section=2&pID=83 &subsection=0 (accessed 10/12/04).

Construction News (2000) 'Imagine there's no site labour'. David Rogers, 7 January, pp. 22–3.

Construction News (2002a) 'Could you face paxman?'. Emma Crates, 7 March, pp. 24–5.

Construction News (2002b) 'Smarten up in Canary Wharf, workers told'. Russell Lynch, 25 April, p. 4.

Construction News (2003) 'Workers get multicultural'. Newsdesk, 30 January, p. 1.

Construction News (2004a) 'The McAlpine family at war'. David Rogers, 11 March, pp. 4–5.

Construction News (2004b) 'It's a family affair'. Andrew Gaved, 8 April, pp. 12–13.

Construction News (2004c) 'Keeping track', in *Construction Products* (supplement), Ranjit Bassi, April/May, pp. 13–16.

Contract Journal (1999a) 'A Millennial Challenge'. CJ2, April (supplement), pp. 4–7.

Contract Journal (2004a) 'Crown Office Drops Bovis Holyrood Case'. 13 October, pp. 4.

Contract Journal (2004b) '12% tested positive for drugs at T5'. 21 January, p. 1.

Contract Journal (2004c) 'Sustainability Losers Protest'. 28 January, p. 4.

Contract Journal (2004d) 'Say it By Text'. 25 February, p. 1.

Cooper, P. (2000) *Building Relationships: The History of Bovis 1885–2000*, Cassell & Co, London.

Cornick, T. and Mather, J. (1999) *Construction Project Teams: Making Them Work Profitably*, Thomas Telford, London.

Cox, I. and Preece, C. (2000) *Using a Website as a Marketing Communications Tool in Construction: Survey into the Effectiveness of the Websites and Back Office Systems of the Top 50 UK Contractors*, Published by Construction Management Group, School of Civil Engineering, Universiy of Leeds, Leeds.

Cox, S. and Hamilton, A. (1991) *Architect's Handbook of Practice Management*, RIBA Publications, London.

Crichton, C. (1966) *Interdependence and Uncertainty – A Study of the Building Industry*, Tavistock Publications, London.

CSSC (1991) *Construction Management Forum, Report and Guidance*, Centre for Strategic Studies in Construction University of Reading, UK.

Culp, G. and Smith, A. (2001) 'Understanding psychological type to improve project team performance'. *Journal of Management in Engineering*. 17(1): 24–33.

CYTGPDC (1999) *Three Gorges Project*, China Yangtze Three Gorges Project Development Corporation.

Dadji, M. (1988) 'Are you talking to M&E?' *RIBA Journal (Switch Supplement)*. October, 102–3.

Dainty, A.R.J. and Moore, D.R. (2000) 'The performance of integrated D&B project teams in unexpected change event management', *ARCOM 2000*, A. Akintoye (ed.), Glasgow, ARCOM Vol. 1, pp. 281–90.

Dainty, A.R.J., Bagilhole, B.M. and Neale, R.H. (2000a) 'A grounded theory of women's career under-achievement in large UK construction companies'. *Construction Management and Economics*. 18: 239–50.

Dainty, A.R.J., Bagilhole, B.M. and Neale, R.H. (2000b) 'The compatibility of construction companies' human resource development policies with employee career expectations'. *Engineering, Construction and Architectural Management*. 7(2): 169–78.

Dainty, A.R.J., Bagilhole, B.M., Ansari, K.H. and Jackson, J. (2002) 'Diversification of the UK construction industry: a framework for change'. *ASCE Journal of Leadership and Management in Engineering*. 2(4): 16–18.

Dainty, A.R.J., Ison, S.G. and Root, D.S. (2004) 'Bridging the skills gap: a regionally driven strategy for resolving the construction labour market crisis'. *Engineering, Construction and Architectural Management*. 11(4): 275–83.

Dainty, A.R.J., Qin, J. and Carrillo, P.M. (2005) 'HRM strategies for promoting knowledge sharing within construction project organizations: a case study', in *Knowledge Management in the Construction Industry: A Socio-Technical Perspective*, A.S. Kazi (ed.), Idea Group, Hershey, PA, pp. 18–33.

Dalton, M. (1959) *Men Who Manage*, John Wiley, New York.

Daly, F., Teague, P. and Kitchen, P. (2003) 'Exploring the role of internal communication during organisational change'. *Corporate Communications: An International Journal*. 8(3): 153–62.

Davidson, F.P. and Huot, J.-C. (1989) 'Management trends for major projects'. *Project Appraisal*. September, 4(3): 137.

Davis, K. (1953) 'Management communication and the grapevine'. *Harvard Business Review*. 4: 43–9.

Davis, K. and Newstorm, J.W. (1985) *Human Behaviour at Work: Organizational Behaviour* (7th Edn), McGraw Hill, New York.

Deal, T.E. and Kennedy, A.A. (1982) *Corporate Cultures: The Rites and Rituals of Corporate Life*, Addison-Wesley, Reading, MA.

Deetz, S. (2001) 'Conceptual foundations', in *The New Handbook of Organizational Communication*, F.M. Jablin and L.L. Putnam (eds), Sage, Thousand Oaks, CA, pp. 3–46.

Delisle, C.L. and Oslon, D. (2004) 'Would the real project management language please stand up?'. *International Journal of Project Management*. 22: 327–37.

Deresky, H. (1994) *International Management: Managing Across Borders and Cultures*, Harper Collins, New York.

D'Herbemont, O., Cesar, B. (1998) *Managing Sensitive Projects*, Macmillan Business Press, London.

Dickson, M. (2003) Modern Construction – Achieving a Step Change in Performance, Presentation to nCRISP Awayday, London, March 11.

Dikmen, I., Birgonul, M.T. and Arikan, A.E. (2004) 'A critical review of risk management support tools', in Proceedings of 20th annual ARCOM Conference, 1–3 September, Heriot Watt University, 2: 1145–54.

Dingle, J. (1997) *Project Management: Orientation for Decision Makers*, Arnold, London.

Dixon, M. (2000) *Association for Project Management Project Management Body of Knowledge* (4th Edn), Association for Project Management, Buckinghamshire.

Dolphin, R.R. (2002) 'A profile of PR directors in British companies'. *Corporate Communications: An International Journal*. 7(1): 17–24.

Dow, G.K. (1988) 'Configurational and co-activational views of organizational structure'. *Academy of Management Review*. 13(1): 53–64.

Dowling, G.R. (1986) 'Managing your corporate image', *Industrial Marketing Management*. 15: 109–15.

Drucker, P.F. (1993) *Post-Capitalist Society*, Oxford: Butterworth Heinemann.

DTI (2004) 'The operating and financial review: practical guidance for directors', www.dti.gov.uk/cld/pdfs/ofr_guide.pdf (accessed 13/01/05).

Dulaimi, M.F. and Dalziel, R.C. (1994) 'Effects of the procurement method on the level of management synergy in construction project', in Proceedings of CIB W92, Procurement Systems East Meets West Symposium, Hong Kong, 4–7 December, 53–9.

Duyshart, B., Mohamed, S., Hampson, K. and Walker, D. (2003) 'Enabling improved business relationships – how information technology makes a difference', in *Procurement Strategies: A Relationship-based Approach*, Walker, Derek and Hampson, Keith (eds), Blackwell Publishing, Oxford, pp. 123–66.

Duyshart, B., Walker, D. and Mohamed, S. (2003) 'An example of developing a business model for information and communication technologies (ICT) adoption on construction projects – the National Museum of Australia project'. *Engineering, Construction and Architectural Management*. 10(3): 179–92.

Edum-Fotwe, F.T., Price, A.D.F. and Thorpe, A. (1996) 'Analysing construction contractors' strategic intent from mission statements', in *The Organisation and Management of Construction: Shaping Theory and Practice – Managing the Construction Enterprise*, Langford, D.A. and Retik, A. (eds), CIB W65, London: Spon, pp. 33–43.

Egan, J. (1998) *Rethinking Construction: The Report of the Construction Task Force*, Department of the Environment, Transport and the Regions, London.

Egbu, C.O. (1999) 'The role of knowledge management and innovation in improving construction competitiveness'. *Building Technology and Management Journal.* 25: 1–10.

Einwiller, S. and Will, M. (2002) 'Towards an integrated approach to corporate branding – an empirical study'. *Corporate Communications: An International Journal.* 7(2): 100–9.

Eisenberg, E.M. and Goodall Jr, H.L. (1993) *Organizational Communication: Balancing Creativity and Constraint*, St Martins Press, New York.

Emmerson, H. (1962) *Survey of Problems Before the Construction Industries: A Report prepared for the Minister of Works*, HMSO, London.

Emmitt, S. and Gorse, C.A. (2003) *Construction Communication*, Blackwell, Oxford.

Emmott, M. and Hutchinson, S. (1998) 'Employee flexibility: threat or promise', in *Human Resource Management: The New Agenda*, Sparrow, P. and Marchington, M. (eds), Pitman, London, pp. 229–43.

Fisher, B.A. (1980) *Small Group Decision Making: Communication and Group Process*, McGraw-Hill, New York.

Fisher, D. (1981) *Communication in Organizations*, West Publishing Company, New York.

Fisher, D. (1993) *Communication in Organizations* (2nd Edn), West Publishing, Eagan, MN.

Fiske, J. (1990) *Introduction to Communication Studies* (2nd Edn), Routledge, London.

Flexibility (2004a) *Organisations and People.* http://www.flexibility.co.uk/flexwork/general/hr1.htm (accessed 05/08/05).

Flexibility (2004b) *The Complete Guide to Flexible Working.* http://www.flexibility.co.uk/Guide/Content/chapter4.pdf (accessed 05/08/05).

Foley, J. and Macmillan, S. (2005) 'Patterns of interaction in construction team meetings', *Co-Design.* 1(1): 19–37. (http://www.tandf.co.uk)

Fox, S., Marsh, L. and Cockerham, G. (2001) 'Design for manufacture: a strategy for successful application to buildings', *Construction Management and Economics.* 19: 493–502.

France, K. (2002) 'The future of information technology in collaborative construction'. *ICE Proceedings: Civil Engineering.* 150 (Special Issue 2): 4–6.

Francis, H. (2002) 'HRM and the beginnings of organizational change'. *Journal of Organizational Change Management.* 16(3): 309–37.

Frankental, P. (2001) 'Corporate social responsibility – a PR invention?'. *Corporate Communications: An International Journal.* 16(1): 18–23.

Fraser (2004) The Holyrood Inquiry, Conducted by the Rt Hon the Lord Fraser of Carmyllie QC. www.holyroodinquiry.org (accessed 05/08/05).

Fryer, B., Egbu, C., Ellis, R. and Gorse, C. (2004) *The Practice of Construction Management* (4th Edn), Blackwell, Oxford.

Furnham, A. (1997) *The Psychology of Behaviour at Work*, Psychology Press, UK.

Gale, A.W. (1992) 'The construction industry's male culture feminise if conflict is to be reduced: the role of education as a gate-keeper to a male construction industry', in *Construction Conflict Management and Resolution*, P. Fenn and R. Gameson (eds), E&FN Spon, London, pp. 416–27.

Gameson, R.N. (1992) *An investigation into the interaction between potential building clients and construction professionals.* PhD Thesis, University of Reading, UK.

Ganah, A., Anumba, C. and Bouchlaghem, N. (2000) 'The use of visualisation to communicate design information to construction sites', in Proceedings of 16th Annual ARCOM Conference, 6–8 September, Glasgow Caledonian University, UK, 2: 833–42.

Gardiner, P.D. (1993) *Conflict analysis in construction project management.* Unpublished PhD thesis, University of Durham, School of Engineering and Computer Science.

Gayeski, D. (1993) *Corporate Communications Management: The Renaissance Communicator in Information-Age Organizations*, Focal Press/Heinneman, Boston, MA.

Goczol, J. and Scoubeau, C. (2003) 'Corporate communication and strategy in the field of projects'. *Corporate Communications: An International Journal.* 8(1): 60–6.

Gorse, C.A. (2002) *Effective interpersonal communication and group interaction during construction management and design meetings.* PhD Thesis, University of Leicester.

Gorse, C.A. and Emmitt, S. (2003) 'Investigating interpersonal communication during construction progress meetings: challenges and opportunities'. *Engineering, Construction and Architectural Management.* 10(4): 234–44.

Gorse, C.A., Emmitt, S. and Lowis, M. (1999) 'Problem solving and appropriate communication medium', in Proceedings of ARCOM, 15th Annual Conference, Liverpool John Moores University, 1: 511–17.

Gottschalk, S. (1999): 'Speed culture: fast strategies in televised commercial ads'. *Qualitative Sociology.* 22: 4.

Gouran, D.S. and Hirokawa, R.Y. (1983) 'The role of communication in decision making groups: a functional perspective', in *Communications in Transition: Issues and Debates in Current research*, M.S. Mander (ed.), Prager, New York, pp. 168–85.

Green, B. (2002) 'A future built on customers', in *Construction News*, 5 September, Top 100, Hill McGlynn Supplement, 2–3.

Green, S.D. and May, S.C. (2003) 'Re-engineering construction: going against the grain'. *Building Research and Information.* 31(2): 97–106.

Guevara, J.M. and Boyer, L.T. (1981) 'Communication problems within construction'. *Journal of the Construction Division*, in Proceedings of the American Society of Civil Engineers, December, 107(4): 551–57.

Hage, J., Aiken, M. and Marrett, C.B. (1971) 'Organization structure and communications'. *American Sociological Review.* 36: 860–71.

Hall, E.T. (1989) *Understanding Cultural Differences*, Intercultural Press, Yarmouth, ME.

Halsey, R. and Margerison, C.J. (1978) 'Managing a large construction site'. *Management Decision.* 16(4): 246–61.

Handy, C. (1993) *Understanding Organizations* (3rd Edn), Penguin, London.

Handy, C. (1999) *Understanding Organizations* (4th Edn), Penguin, London.

Hargie, O. (1986) *A Handbook of Communication Skills*, Routledge, London.

Harris, S. and Kington, A. (2002) Innovative Classroom Practice using ICT in England: The Second Information in Technology Study (SITES). http://www.nfer.ac.uk/researchareas/pims-data/summaries/ite-the-use-of-ict-in-the- (accessed classroom.cfm (accessed 01/09/05).

Harrower, J. (2003) 'Designer contractors' key to project delivery'. *Contract Journal*. 19 March, 4.

Hawes, L.C. (1974) 'Social collectivies as communication: perspective on organizational behaviour'. *Quarterly Journal of Speech*. 60: 497–502.

Hawes, L.C. (1990) 'Social collectives as communication', in *Foundations of Organizational Communication: A Reader*, Corman, S.R., Banks, S.P., Bantz, C.R., and Mayer, M.E. (eds), Longman, New York, pp. 41–4.

HBG (2001) A Practical Guide to Best Practice for Site based Responsible Community Relations, HBG UK Ltd.

HBG (2004a) Considerate Constructors Scheme Report, HBG UK Ltd.

HBG (2004b) Corporate Social Responsibility Policy, HBG UK Ltd.

Hellard, R.B. (1995) *Project Partnering: Principle and Practice*, Thomas Telford, London.

Henderson, L.S. (1987) 'The contextual nature of interpersonal communication in management theory and research'. *Management Communication Quarterly I*. 1: 7–31.

Henderson, L.S. (2004) 'Encoding and decoding communication competencies in project management – an exploratory study'. *International Journal of Project Management*. 22(6): 469–76.

Higgin, G. and Jessop, N. (1965) *Communication in the Building Industry: The Report Of A Pilot Study*, Tavistock Publications, London.

Hill, C.J. (1995) 'Communication on construction sites', in Proceedings of ARCOM 1995, 18–20 September, University of York, 1: 232–40.

Hillebrandt, P.M. and Cannon, J. (1990) *The Modern Construction Firm*, Macmillan, Hampshire.

Hindle, R.D. and Muller, M.H. (1996) 'The role of education as an agent of change: a two fold effect'. *Journal of Construction Procurement*. 3(1): 56–66.

Hirokawa, R.Y. (1980) 'A comparative analysis of communication patterns within effective and ineffective decision-making groups'. *Communication Monographs*. November, 47: 312–21.

HM Treasury (1992) HM treasury guidance no 36: contract selection for major projects, June.

Holloman, C.R. and Hendrick, H.W. (1972) 'Adequacy of group decisions as a function of the decision-making process'. *Academy of Management Journal*. 15(2): 175–84.

Hopper, J.R. (1990) *Human Factors of Project Organization*, Construction Industry Institute, Austin, TX, Source Document 58.

Howes, R. and Tah, J.H.M. (2003) *Strategic Management Applied to International Construction*, Thomas Telford, London.

HSBC (2004) Business Connections Turkey, East Meets West.

Huber, G. (1990) 'A theory of the effects of advanced information technologies on organizational design, intelligence and decision-making', in *Organizations and Communication Technology*, Fulk, J. and Steinfeld, C. (eds), Sage, CA, 237–74.

Huczynski, A. and Buchanan, D. (2001) *Organizational Behaviour: An Introductory Text* (4th Edn), Essex, Prentice Hall, NJ.

Hughes, W. (1989) *Organizational analysis of building projects*. PhD Thesis, Liverpool John Moores University.

Ikeda, N. (No date) Beyond the Internet. http://www.telecomvisions.com/articles/beyondip/ (accessed 05/08/05).

Ingram, H., Teare, R., Scheuing, E. and Armistead, C. (1997) 'A system model of effective teamwork'. *TQM Magazine*. 9(2): 118–27.

Jackson, P.C. (1987) *Corporate Communication for Managers*, Pitman Publishing, London.

Janis, I.L. (1972) *Groupthink: A Psychological Study of Foreign-Policy Decisions and Fiascoes*, Houghton Mifflin Company, Boston, MA.

Jarvenpaa, S.L. and Leidner, D.E. (1998) 'Communication and trust in global virtual teams'. *Journal of Computer Mediated Communication*. 3(4): 1–38. [Online] http://hyperion.math.upatras.gr/commorg/jarvenpaa/ (accessed 05/08/05).

Johnson, R.E. and Clayton, M.J. (1998) 'The impact of information technology in design and construction'. *Automation in Construction*. 8: 3–14.

Josephson, P.E. and Larsson, B. (2001) 'The role of early detection of human errors in building projects', in Proceedings of CIB World Building Conference, Performance in Product and Practice 2–6 April, Wellington, New Zealand, Paper: HPT 08 (from CD rom).

Kakabadase, A., Ludlow, R. and Vinnicombe, S. (1988) *Working in Organizations*, Penguin, London.

Kamara, J.M., Anumba, C.J. and Evbuomwan, N.F.O. (2000) 'Establishing and processing client requirements – a key aspect of concurrent engineering'. *Construction, Engineering Construction and Architectural Management*. 7(1): 15–28.

Kane, H. (2003) 'Reframing the knowledge debate with a little help from the Greeks'. *Electronic Journal of Knowledge Management*. 1(1): 33–8.

Karger, D.W. and Murdick, R.G. (1963) *Managing Engineering and Research*, Industrial Press Inc., New York.

Katz, D. and Khan, R.L. (1978) *The Social Psychology or Organizations* (2nd Edn), John Wiley, New York.

Keegan, A. and Turner, J.R. (2003) 'Managing human resources in the project-based organization', in *People in Project Management*, Turner, J.R. (ed.), Gower, Aldershot, pp. 1–11.

Keel, D., Hartley-Jones, S. and Ellis, R. (2004) 'Careers in construction: children have their Say!', in Proceedings of RICS COBRA Conference 7–8 September, Leeds Metropolitan University, www.rics-foundation.org/publish/document.aspx?did=3280 (accessed 10/01/05).

Kerzner, H. (1997) *Project Management: A Systems Approach to Planning, Scheduling and Controlling (6th edn)*, Van Nostrand Reinhold Company, New York.

Kier (2005) Kids Zone, www.kier.co.uk/kids/default_flash.asp (accessed 17/12/04).

Kiersey, D. and Bates, M. (1984) *Please Understand Me – Character and Temperament Types* (5th Edn), Prometheus Nemesis Book Company, Delmar, CA.

Kiesler, S., Siegel, J. and McGuire, T.W. (1984) 'Social psychological aspects of computer-mediated communication'. *American Psychologist*. 39: 10.

Kitchen, P. (1997) *Public Relations: Principle and Practice*, Thomson Business Press, London.

Klauss, R. and Bass, B.M. (1982) *Interpersonal Communication in Organizations*, Academic Press, New York.

Knapp, M.L. and Hall, J.A. (2002) *Nonverbal Communication in Human Interaction* (5th Edn), Wadsworth, Ohio, OH.

Krackhardt, D. and Hanson, J.R. (1993) 'Informal networks: the company behind the chart'. *Harvard Business Review*. July/August, 71(4): 104–11.

Kreps, G.L. (1989) *Organizational Communication: Theory and Practice* (2nd Edn), Longman, New York.

Lancaster, R.J., McAllister, I. and Alder, A. (2001) *Establishing Effective Communication and Participation in the Construction Sector*, Report prepared by Entec UK Ltd for the Health and safety Executive, HSE Books, Suffolk, UK.

Langford, D., Hancock, M.R., Fellows, R. and Gale, A.W. (1995), *Human Resources Management in Construction*, Longman, Essex.

Latham, M. (1994) *Constructing the Team*, HMSO, London.

Laudon, K.C. and Laudon, J.P. (2002) *Management Information Systems: Managing the Digital Firm*, Prentice Hall, NJ.

Laufer, A., Woodward, H. and Howell, G.A. (1999) 'Managing the decision-making processes during project planning'. *Journal of Management in Engineering*. 15(2): 79–84.

Lewis, D. (2004) 'Communication breakdown Key to Fiasco says Holyrood Engineer'. *Architects Journal*. 13 May, 13.

Lipnack, J. and Stamps, J. (1997) *Virtual Teams: Reaching Across Space, Time and Organisations with Technology*, John Wiley and Sons, New York.

Loosemore, M. (1996) *Crisis management in building projects: a longitudinal investigation of communication behaviour and patterns within a grounded framework*. Unpublished PhD Thesis, University of Reading, Berkshire.

Loosemore, M. (2000) *Construction Crisis Management*, American Society of Civil Engineers, New York.

Loosemore, M. and Al Muslmani, H.S. (1999) 'Construction project management in The Persian Gulf – inter-cultural communication'. *International Journal of Project Management*. 17(2): 95–101.

Loosemore, M. and Lee, P. (2002) 'Communication problems with ethnic minorities in the construction industry'. *International Journal of Project Management*. 20(7): 517–24.

Loosemore, M. and Tan, C.C. (2000) 'Occupational bias in construction management research'. *Construction Management and Economics*. 18(7): 757–66.

Loosemore, M., Dainty, A.R.J. and Lingard, H. (2003) *HRM in Construction Projects: Strategic and Operational Approaches*, E&FN Spon Press, London.

Loosemore, M., Nguyen, B.T. and Denis, N. (2000) 'An investigation into the merits of encouraging conflict in the construction industry'. *Construction Management and Economics*. 18(4): 447–56.

Love, P.E.D. and Li, H. (2000) 'Quantifying the causes and costs of rework in construction'. *Construction Management and Economics*. 18: 479–90.

Love, P.E.D., Li, H., Tse, R.Y.C. and Cheng, E.W.L. (2000) 'An empirical analysis of IT/IS evaluation in construction'. *The International Journal of Construction Information Technology*. 8(2): 21–38.

Lurey, J.S. and Raisinghani, M.S. (2001) 'An empirical study of best practices in virtual teams'. *Information and Management*. 38: 523–44.

McDermott, R. and O'Dell, C. (2001) 'Overcoming cultural barriers to sharing knowledge'. *Journal of Knowledge Management*. 5(1): 76–85.

McLure Wasko, M. and Faraji, S. (2000) 'It is what one does: why people participate and help others in electronic communities of practice'. *Journal of Strategic Information Systems.* 9: 155–73.

McMurdo, G. (2004) Changing Contexts of Communication. http://jimmy.qmuc. ac.uk/jisew/ewv21n2/ (accessed 05/08/05).

Malone, T.W. and Smith, S.A. (1984) Tradeoffs in designing organisations: implications for new forms of human organisations and computer systems. Working Paper 112, Center for Information Systems Research. MIT.

Marchington, M. and Grugulis, I. (2000) 'Best practice human resource management: perfect opportunity or dangerous illusion?'. *International Journal of Human Resource Management.* 11(4): 905–25.

Maurer, J.G. (1992) 'Foreword', in *Organization Charts*, Nixon, J.M. (ed.), Gale Research Inc, London, p.vii.

Mayo, E. (1945) *The Social Problems of an Industrial Civilization*, New Hampshire: Ayer.

Melvin, T. (1979) *Practical Psychology in Construction Management*, Van Nostrand Reinhold, New York.

Metts, S. and Bowers, J.W. (1994) 'Emotions in interpersonal communications', in *Handbook of Interpersonal Communication* (2nd Edn), Knapp, M.L. and Miller, G.R. (eds), SAGE publications, London, pp. 508–41.

Miller, R. and Lessard, D.R. (2000) *The Strategic Management of Large Engineering Projects: Shaping Institutions, Risks, and Governance*, Massachusetts Institute of Technology, Cambridge, MA.

Mintzberg, H., Raisinghani, D. and Theorot, A. (1976) 'The Structure of Unstructured Decision Process', *Administrative Science Quarterly*, 21: 246–274.

Monge, P.R. and Contractor, N.S. (1988) 'A handbook for the study of human communication: methods and instruments for observing, measuring, and assessing communication processes', in *Communication Networks: Measurement Techniques*, Tardy, C.H. (ed.), Ablex Publishing, Norwood, NJ, pp. 107–38.

Moore, D.R. (1996) 'The renaissance: the beginning of the end for implicit buildability'. *Building Research Information.* 24(5): 259–69.

Moore, D.R. (2001) 'William of Sen to Bob the builder: non-cognate cultural perceptions of constructors'. *Engineering, Construction and Architectural Management.* 8(3): 177–84.

Moore, D.R. (2002) *Project Management: Designing Effective Organisational Structures in Construction*, Blackwell Science, Oxford.

Moore, D.R. and Abadi, M. (2005) 'Virtual teamworking and associated technologies within the UK construction industry'. *Architecture, Engineering & Design Management.* 1(1).

Moore, D.R. and Dainty, A.R.J. (1999) 'Integrated project teams' performance in managing unexpected change events'. *Team Performance Management.* 5(7): 212–22.

Moore, D.R. and Dainty, A.R.J. (2000) 'Work-group communication problems within UK design and build projects: an investigative framework'. *Journal of Construction Procurement*, May 6(1): 44–55.

Moore, D.R. and Dainty, A.R.J. (2001) 'Intra-team boundaries as an inhibitor to performance improvement in the UK construction industry'. *Construction Management and Economics.* 19(6): 559–62.

Moore, D.R., Cheng, M.-I. and Dainty, A.R.J. (2002) 'Competence, competency and competencies: performance assessment in organisations'. *Work Study: A Journal of Productivity Science.* 51(6): 314–19.

Moreland, R.L. and Levine, J.M. (2002) 'Socialization and trust in work groups'. *Group Processes and Intergroup Relations.* 5(3): 185–202.

Mott Macdonald (2002) Review of Large Public Procurement in the UK, A Report commissioned by HM Treasury.

Mouritsen, J. and Bjorn-Andersen, N. (1991) 'Understanding third-wave information systems', in *Computerisation and Controversy: Value Conflicts and Social Choices*, Dunlop, C. and King, R. (eds), Academic Press, California, pp. 308–20.

Moxley, R. (1991) 'Motivating professionals', in *Practice Management: New Perspectives for the Construction Professional*, Barrett, P. and Males, R. (eds), E&FN Spon, London, pp. 304–13.

Moxley, R. (1993) *Building Management by Professionals*, Butterworth Architectural Management Guides, Oxford.

Muller, A. (1997) *Radix: Matrix: Works and Writings of Daniel Liebskind.* Prestel Verlag, Munich.

Mullins, L.J. (1999) *Management and Organisational Behaviour* (5th Edn), Pearson Education, Essex.

Murdoch, A. (1997) 'Human re-engineering', *Management Today*, 6–9.

Murray, M. (2003a) 'Rethinking construction: the egan report (1998)', in *Construction Reports 1944–98*, Murray, M. and Langford, D. (eds), Blackwell Publishing, Oxford, pp. 173–95.

Murray, M. (2003b) *Building perceptions with metaphors: a study of the communication and decision-making behaviour of construction professionals within projects.* Unpublished PhD Thesis, University of Strathclyde, Glasgow.

Murray, M. and Langford, D. (2003) *Construction Reports 1944–98*, Blackwell Publishing, Oxford.

Murray, P. (2004) *The Saga of Sydney Opera House*, Spon Press, London.

Musgrave, E.C. (1994) 'The organisation of the building trades of Eastern Brittany 1600–1790: some observations'. *Construction History.* 10: 1–13.

National Audit Office (2001) Modernising Construction, Report by the Controller and Auditor General, London.

National Construction Week (2004) www.ncw.org.uk/index.cfm (accessed 28/06/04).

NEDO (1988) National Economic Development Office, Faster Building for Commerce, HMSO, London.

Nesan, L.J. and Holt, G.D. (1999) *Empowerment in Construction; The Way Forward for Performance Improvement*, Research Studies Press Ltd, London.

Ng, S.T., Chen, S.E., McGeorge, D., Lam, K.C. and Evans, S. (2001) 'Current state of IT usage by Australian sub-contractors'. *Construction Innovation.* 1(1): 3–13.

Nicolini, D. (2001) In Search of 'Project Chemistry': An Initial Review of Project Issues and their Impact on the Performance of Construction Projects, CRISP Commission 00/9, The Tavistock Institute.

Nicolini, D. (2002) 'In search of project chemistry'. *Construction Management and Economics.* 20: 167–77.

Nonaka, I. and Takeuchi, H. (1995) *The Knowledge Creating Company*, Oxford University Press, Oxford.

Norton, R.W. (1978) 'Foundation of a communicator style construct'. *Human Communication Research*. 4: 99–112.

O'Cathain, C. and Gallacher, S. (1999) An analysis of breakdowns in communication between design and project implementation, International Symposium of design Science, 4th Asian design Conference, 30–31 October. http://143.117.118.42/Cladding/4thADC/Japan99.htm (accessed 30/08/00).

Pacanowsky, M.E. and O'Donnell-Trujillo, N. (1990) 'Communication and organizational cultures', in *Foundations of Organizational Communication: A Reader*, Corman, S.R., Banks, S.P., Bantz, C.R. and Mayer, M.E. (eds), Longman, New York, pp. 142–53.

Partington, D. (2003) 'Managing and leading', in *People in Project Management*, J.R. Turner (ed.), Gower, Aldershot, pp. 83–97.

Pearce, D. (2003) *The Social and Economic value of Construction: The Construction Industry's Contribution to Sustainable Development*, New Construction Research and Innovation Strategy Panel, London.

Pennington, D.C. (1986) *Essential Social Psychology*, Arnold, London.

Pettit, J.D. Jr, Goris, J.R. and Vaught, B.C. (1997) 'An examination of organizational communication as a moderator of the relationship between job performance and job satisfaction'. *Journal of Business Communication*. 34(1): 8–26.

Pinto, M.B. and Pinto, J.K. (1991) 'Determinants of cross-functional cooperation in the project implementation process'. *Project Management Journal*. 22: 13–20.

Polanyi, M. (1966) *The Tacit Dimension*, Routledge & Kegan Paul, London.

Poole, M.S. and Desanctis, G. (1990) 'Understanding the use of group decision support systems: the theory of adaptive structuration', in *Organizations and Communication Technology*, Fulk, J. and Steinfeld, C. (eds), Sage, Thousand Oaks, CA, pp. 173–93.

Preece, C.N., Moodley, K. and Smith, A.S. (1998) *Corporate Communication in Construction: Public Relations Strategies for Successful Business and Projects*, Blackwell Science, Oxford.

Preece, C., Moodley, K. and Cox, I. (2001) 'Assessing the effectiveness of websites as an interactive business communications tool', in *Seventeenth Annual ARCOM Conference*, 5–7 September 2001, Akintoye, A. (ed.), University of Salford. Association of Researchers in Construction Management, 1, 207–17.

Project Management Institute (PMI) (2000) *A Guide to the Project Management Body of Knowledge*, Project Management Institute, Newtown Square, PA.

Quirke, B. (1995) *Communicating Corporate Change*, McGraw-Hill, New York.

Raghuram, S., Raghu, G., Wiesenfeld, B. and Gupta, V. (2001) 'Factors contributing to virtual work adjustment'. *Journal of Management*. 27: 383–405.

Reid, J.L. (2000) *Crisis Management: Planning and Media Relations for the Design and Construction Industry*, John Wiley & Sons, New York.

Reimer, J.W. (1979) *Hard Hats: The Work World of Construction Workers*, Sage Publications, Beverly Hills, CA.

Retik, A. and Langford, D. (2001) *Computer Integrated Planning and Design for Construction*, Thomas Telford, London.

Reuters (2004) China Halts Dam, Sacks Official After Protests. Reuters Foundation, 18 November 2004. www.alertnet.org/thenews/newsdesk/PEK160131.htm

RIBA (1990) *RIBA Survey of Computer Usage 1989*, RIBA, London.

Richmond, V.P., McCroskey, J.C. and Payne, S.K. (1991) *Nonverbal Behaviour in Interpersonal Relations* (2nd Edn), Prentice Hall, Englewood Cliffs, NJ.

Riemer, J.W. (1979) *Hard Hats: The Work World of Construction Workers*, Sage Publications, Beverly Hills, CA.

Riggenbach, J.A. (1986) 'Silent negotiation: listen with your eyes'. *Journal of Management in Engineering.* 2(2): 91–100.

Riley, M. (2004) Managing Risks, Quoted in New Civil Engineering Terminal 5 Supplement, February 2004, pp. 22–3.

Roberts, C., Edwards, R. and Barker, L. (1987) *Intrapersonal Communication Processes*, Gorsuch Scarisbrick, Scottsdale, AZ.

Roberts, K.H. and O'Reilly, C.A. (1978) 'Organizations as communications structures: an empirical approach'. *Human Communication Research.* 4: 283–93.

Rogers, E.M. and Agarwala-Rogers, E. (1976) *Communication in Organizations*, The Free Press, London.

Rooke, J., Seymour, D. and Fellows, R. (2003) 'The claims culture: a taxonomy of attitudes in the industry'. *Construction Management and Economics.* 21(2): 167–74.

Rosenthal, M.J. (2001) 'High-performance teams'. *Executive Excellence.* 18(10): 6.

Ryan, H. (2001) 'Image is everything'. *Building.* 4 May, 52–4.

Scarborough, H. (1999) 'System error'. *People Management.* 8 April, 68–74.

Schein, E.H. (1985) *Organizational Culture and Leadership: A Dynamic View*, Jossey-Bass, San Francisco, CA.

Schellekens, P. and Smith, J. (2004) *Language in the Construction Industry: Communicating with second language speakers*, Report by the Schellekens Consultancy, London.

Schwartz, D.F. and Jacobson, E. (1977) 'Organizational communication network analysis: the liason communication role'. *Organizational Behaviour and Human Performance.* 18: 158–74.

Schweiger, D. and De Nisi, A. (1991) 'Communicating with employees following a merger: a longitudinal field experiment'. *Academy of Management Journal.* 34(1): 100–35.

Scott, W.R. (1981) *Organizations: Rational, Natural and Open Systems.* Prentice Hall, Englewood Cliffs, NJ.

Senge, P., Kleiner, A., Roberts, C., Ross, R., Roth, G. and Smith, B. (1999) *The Dance of Change: A fifth Discipline Resource*, Nicolas Brealey, London.

Shannon, C.E. and Weaver, W. (1949) *The Mathematical Theory of Communication.* University of Illinois, Urbana, IL.

Sheldrick-Ross, C. and Dewdney, P. (1998) *Communicating Professionally* (2nd Edn), Library Association Publishing, London.

Shohet, I.M. and Frdman, S. (2003) 'Communication patterns in construction at construction manager level'. *Journal of Construction Engineering and Management.* ASCE, September/October 2003, 570–77.

Sidwell, A.C. (1990) 'Project management: dynamics and performance'. *Construction Management and Economics.* 8: 159–78.

Silverman, D. (1970) *The Theory of Organisations*, Heinemann, London.

Skyttner, L. (1998) 'Some complementary concepts of communication theory'. *Kybernetics: The International Journal of Systems & Cybernetics.* 27(2): 155–64.

Smallman, C. and Weir, D. (1999) 'Communication and cultural distortion during crises'. *Disaster Prevention and Management.* 8(1): 33–41.

Smith, E. (2001) 'The role of tacit and explicit knowledge in the workplace'. *Journal of Knowledge Management.* 5(4): 311–21.

Stallworthy, E.A. and Khardanda, O.P. (1985) *International Construction and the Role of the Project Manager,* Gower, London.

Stevens, J.E. (1988) *Hoover Dam: An American Adventure,* University of Oklahoma press, Norman, OK.

Stokes, E. (2004) *ICT Learning and Training in Ireland – Policies and Data,* Centre for Educational Research, London School of Economics and Political Science. February.

Storey, J. (1993) The take-up of human resource management by mainstream companies: key lessons from research. *The International Journal of Human Resource Management.* 4(3): 529–53.

Strategic Forum (2002) *Accelerating Change,* Strategic Forum for Construction, London.

Swan, J., Newell, S. and Robertson, M. (2000) 'Knowledge management – when will people management enter the debate?', in Proceedings of the 33rd Hawaii International Conference on System Sciences, 4–7 January, Maui, Hawaii, 3.

Talk: wildlife (2003) Baiji Dolphin may not survive another decade! http://www.talkwildlife.citymax.com/page/page/532558.htm

Tam, C.M., Fung, I.W.H., Yeung, T.C.L. and Tung, K.C.F. (2003) 'Relationship between construction safety signs and symbols recognition and characteristics of construction personnel'. *Construction Management and Economics.* 21: 745–53.

The Telegraph (2001) BA restores Union Flag Design to All Tailfins. www.telegraph. co.uk/news/main.jhtml?xml=/news/2001/05/11/nfin11.xml (accessed 15/06/04).

Thomas, S.R., Tucker, R.L. and Kelly, R. (1998) 'Critical communication variables'. *Journal of Construction Engineering and Management.* 124(1): 58–66.

Thomason, G. (1988) *A Textbook of Human Resource Management,* Institute of Personnel management, London.

Thompson, M., Ellis, E. and Wildavsky, A. (1990) *Cultural Theory,* Westview Press, Oxford.

Thompson, P. and Mchugh, D. (2002) *Work Organisations: A Critical Introduction* (3rd Edn), Palgrave, Hampshire.

Thorpe, A. and Mead, S. (2001) 'Project-specific web sites: friend or foe'. *Journal of Construction Engineering and Management.* September/October, 127(5): 406–13.

Torrington, D. and Hall, L. (1998) *Human Resource Management* (4th Edn), Prentice Hall, London.

Torrington, D., Weightman, J. and Johns, K. (1995) *Management Methods,* IPM, London.

Townly, B. (1994) 'Communicating with employees', in *Personnel Management: A Comprehensive Guide to Theory and Practice in Britain,* Sisson, K. (ed.), Blackwell, Oxford, pp. 595–633.

Townsend, A.M., DeMarie, S.M. and Hendrickson, A.R. (1998). 'Virtual teams: Technology and the workplace of the future'. *The Academy of Management Executive.* 12(3): 17–29.

Tucker, R.L., Kelly, W.R. and Thomas, S.R. (1997) An Assessment Tool for Improving Team Communications, Construction Institute Report 105–11, Austin, TX.

Tuckman, B.W. (1965) 'Developmental sequences in small groups'. *Psychological Bulletin*. 63: 384–99.

Turner, B.A. and Pidgeon, N.F. (1997) *Man-made Disasters* (2nd Edn), Butterworth, Heinemann, Oxford.

Turner, J.R. (1998) *The Handbook of Project-based Management* (2nd Edn), McGraw-Hill, London.

Turner, J.R. and Muller, R. (2003) 'On the nature of the project as a temporary organisation'. *International Journal of Project Management*. 21: 1–8.

van Raaij. W.F. (1986) Impressive management: het communicatiebleid van de Onderneming, Tekst uitgesproken tijdens de Industriele Communicatiedag van de Bond van Adverteerders, Eramus Universiteit, Rotterdam.

van Rekom, J.V., van Riel, C.B.M. and Wierenga, B. (1991) Corporate Identity. van vagg concept narr hard feitenmateriaal, working paper, Corporate Communication Centre; Erasmus University Rotterdam.

van-Riel, C.B.M. (1995) *Principles of Corporate Communication*, Prentice Hall, London.

Virilio, P. (1995) 'Speed and information: Cyberspace alarm'. *CTheory*. World Wide Web 18 (3) Article 30, http://epe.lacbac.ga.ca/100/201/300/ctheory/articles/1995/art30.txt, (accessed 1/5/05).

Walker, A. (1980) *A model of the design of project management structures for building clients*. Unpublished PhD Thesis, Liverpool Polytechnic, Liverpool, UK.

Walker, A. (2002) *Project Management in Construction*, Blackwell Science, Oxford.

Wantanakorn, D., Mawdesley, M.J. and Askew, W.H. (1999) 'Management errors in construction, engineering construction and architectural management'. *Construction, Engineering Construction and Architectural Management*, 6(2): 112–20.

Wearne, P. (1999) *Collapse: Why Buildings Fall Down, Channel Four Books*, Macmillan Publishers Ltd, London.

Weick, K.E. (1987) 'Theorising about organizational communication', in *Handbook of Organizational Communication: An Interdisciplinary Perspective*, Jublin, F.M., Putnam, L.L., Roberts, K.H. and Porter, L.W. (eds), Sage Publications, Newbury Park, CA, pp. 97–122.

Weinshall, T.D. (1979) *Managerial Communication: Groups, Approaches and Techniques*, Academic Press, A Subsidiary of Harcourt Brace Jovanovisch, Publishers, London.

Weippert, A., Kajewski, S.L. and Tilley, P.A. (2003) 'The implementations of online information and communication technology (ICT) on remote construction sites'. *Logistics Information Management*. 16(5): 327–40.

Weiss, D.S. (2000) *High Performance HR: Leveraging Human Resources for Competitive Advantage*. John Wiley & Sons, Ontario.

Wenger, E. (1998) *Communities of Practice. Learning, Meaning and Identity*. Cambridge University Press, Cambridge.

Wenger, E., McDermott, R. and Snyder, W.M. (2002) *Cultivating Communities of Practice*, Harvard Business School Press, Boston, MA.

Whyte, J., Bouchlaghem, N.M. and Thorpe, A. (2002) 'IT implementation in the construction organization'. *Engineering, Construction and Architectural Management*. 9(5/6): 371–7.

Wilkinson, A. (2001), 'Empowerment', in *Contemporary Human Resource Management*, Redman, T. and Wilkinson, A. (eds), Pearson, Harlow, pp. 336–52.

Williams, J.C. (1988) 'A human factors database to influence safety and reliability. Human factors and decision making: their influence on safety and reliability', in *Symposium for the Safety and Reliability Society*, Sayers, B.A. (ed.), pp. 223–40.

Williams, R. (2004) Marketing View: Non-Compliance, taken from www.housebuilder.co.uk/articles/printable.php?id=1494 (accessed 31/10/2004).

Wilson, D. (2000) 'Metarepresentation in linguistic communication', in *Metarepresentations*, Sperber, D. (ed.), Oxford University Press, Oxford, pp. 411–48.

Wofford, J.C., Gerloff, E.A. and Cummins, R.C. (1977) *Organizational Communication: The Keystone to Managerial Effectiveness*, McGraw-Hill, New York.

Wogalter M.S. and Sojourner, R.J. (1997) 'Comprehension and retention of safety pictorials'. *Ergonomics*. 40(5): 531–42. (http://www.tandf.co.uk)

Wolf Olins (2004) Bovis: Brilliant Precision, taken from www.wolff-olins.com/files/BovisCase Study.pdf (accessed 31/10/2004).

Wombat (1997) 'Moore's law'. *Free on Line Dictionary of Computing*. http://wombat.doc.ic.ac.uk/foldaoc/foldoc.cgi?Moore's+Law (accessed 17/03/2005).

Wong, C.H. and Sloan, N. (2004) 'Use of ICT for e-procurement in the UK construction industry: a survey of SMEs readiness', in Proceedings of the 20th Annual ARCOM Conference, Khosrowshahi, F. (ed.), Edinburgh, 1: 620–8.

World Wildlife Fund (2003) Building Towards Sustainability: Performance and Progress Amongst the UK's Leading Housebuilders. www.wwf.org.uk/filelibrary/pdf/bts.pdf (accessed 10/03/05).

Xybernaut (2004) 'CSX. Upgrading event reporting to real-time'. *Transportation Case Study*. www.xybernaout.com (accessed 21/03/2005).

Index